AF340681

Protein and Peptide Analysis by LC-MS
Experimental Strategies

Protein and Peptide Analysis by LC-MS
Experimental Strategies

Edited by

Thomas Letzel
*Competence Pool Weihenstephan, Technische Universität München,
85354 Freising - Weihenstephan, Germany*

RSC Publishing

RSC Chromatography Monographs No. 15

ISBN: 978-1-84973-182-9
ISSN: 1757-7055

A catalogue record for this book is available from the British Library

Published by The Royal Society of Chemistry,
Thomas Graham House, Science Park, Milton Road,
Cambridge CB4 0WF, UK

Registered Charity Number 207890

For further information see our web site at www.rsc.org

Printed in Great Britain by CPI Antony Rowe, Chippenham and Eastbourne

Foreword

Modern analysis of protein molecules reflects the highly processed nature of these molecules in biological systems. There is a gradient from DNA to RNA to proteins, not just through the 'central dogma' of information flow in evolved biopolymers, but also in terms of complexity. DNA contains some minor variation (polymorphisms), RNA molecules far more, *via* splicing and editing, that creates a far greater number of transcripts than genes in eukaryotic systems from yeast to mammals. Proteins are perhaps the most complicated biopolymers (includes glycoproteins) and yet technologies are not generally considered comprehensive like their PCR-enabled counterparts in functional genomics. Therefore, and particularly for this dynamic area, researchers must stay abreast of the latest in developments, as great strides in data collection and data processing are being realized as the second 'postgenomic' decade dawns.

This collection of 12 chapters provides a one-stop reference for diverse and pragmatic aspects of this evolving area of measurement science. General strategies of 'top down' and 'bottom up' are described in Chapter 1, along with targeted strategies for detection of modifications and quantitation over the next several chapters. Each of the three main areas of the generic workflow is considered: sample preparation with a heavy dose of chromatography, the nuts and bolts of high- and low-performance mass spectrometry, and several descriptions of software that can help in converting raw data into knowledge. Several timely trends in the field are captured in this volume, such as the push toward handling ever-larger proteins as intact species prior to LC-MS, confident assignment of post-translational modifications, and use of tandem mass spectrometry with high mass accuracy.

This volume captures the state-of-the-art for proteomics aficionados, yet strikes a good balance to assist the entry-level cell biologist or protein chemist with specific protocols. The contributions are international in flavour, with chapters from authors in three continents including Japan, the United States, and Europe. Diverse readers should look forward to a spectacular next

RSC Chromatography Monographs No. 15
Protein and Peptide Analysis by LC-MS: Experimental Strategies
Edited by Thomas Letzel
© The Royal Society of Chemistry 2011
Published by the Royal Society of Chemistry, www.rsc.org

half-decade that will be unlike any other in this interdisciplinary area. Thus I hope this volume will solidify core competencies in protein mass spectrometry for measurement scientists and those outside proteomics who want in—for there is plenty of room in the sandbox.

Neil L. Kelleher
Evanston, IL

Preface

Ten years ago, – working as an analytical chemist with environmental aerosols – I could not imagine that I would one day edit a book about the analysis of proteins and peptides. However, studying small molecules at that time with GC-MS and using LC-MS systems for the first time triggered me to develop new analytical techniques and to apply them to versatile types of organic molecules. Meanwhile I applied various LC-MS systems for almost each type of molecules (from hydrophilic to hydrophobic, from small to large, from acidic to basic, from anthropogenic to biological sources, from air to wine). Thereby I realized that it is essential to understand both: the molecules and the analytical systems. Unfortunately this experience is hard to get. But even harder to obtain is written material for teaching the coherence and practical needs of both.

Every molecular biologist intensively learns how to work with techniques like gel electrophoresis or chromatographic purification of proteins, but only few biological researchers learn to handle protein analysis by means of state-of-the-art LC-MS. One reason seems to be the complex instrumentation obviously requiring an experienced operator. Another reason possibly is the enormous speed in the development of new LC-MS systems. In the last ten years, these systems reached a half-life like their embedded computers (something between half a year and one year). Half-life thereby does not mean the working period but the time point when to find the next MS generation on market. However, each new generation has an immense influence on the analytical response, like sensitivity, specifity, and speed. Simultaneously, the LC-MS world became more complicated; for example, instead of just LC there are RP-HPLC, nanoLC and UHPLC; instead of one ionization source there are ESI, APCI and APPI; instead of a tandem mass spectrometer there are QToF, TripleQ and Orbitrap. Even in protein identification and quantification strategies it is not easier, SILAC, ICAT, iTRAQ, TMT and there still is no end in sight.

RSC Chromatography Monographs No. 15
Protein and Peptide Analysis by LC-MS: Experimental Strategies
Edited by Thomas Letzel
© The Royal Society of Chemistry 2011
Published by the Royal Society of Chemistry, www.rsc.org

Obviously, this book cannot reflect the entire variations and applications of LC-MS techniques in protein and peptide analysis and while not attempting that, the book does provide a professional introduction to this popular topic. Moreover it also should give experienced experts one or more tips and hints about special handling procedures.

I am very proud that I could include book chapters from several experts on-the-job. They are presenting their LC-MS strategies for the analysis of peptides and proteins with really detailed information about the needs and problems in their daily work.

Chapter 1 starts with a critical discussion about the current status of 'bottom-up' and 'top-down' strategies in a quantitative point of view. The author Friedrich Lottspeich knows very well about the strengths and weaknesses of LC-MS in protein analysis since he is a proteomics researcher from the early days.

Chapter 2 gives a short introduction into the nomenclature of analytical LC-API-MS techniques and possibilities of coupling several instrumental parts. Hereby, a key aspect is laid on the atmospheric pressure ionization (API) typically neglected in the abbreviation LC-MS and its descriptions.

Chapter 3 is written by Japanese colleagues Tamo Fukamizo and Takayuki Ohnuma representing a laboratory that expresses and purifies protein in high quality and with a lot of experience. The chapter gives experimental insights how to apply highly sophisticated chromatography for protein purification.

Chapter 4 – by American colleagues Nicolas L. Young and Benjamin A. Garcia – continues the chromatography of proteins to an effective hyphenation with mass spectrometry. They impressively describe the does and don'ts with high intuition for the molecules and analytical systems.

All subsequent chapters have European authors reflecting the strength of European protein analysis but, to some degree, my personal contacts and network as well. None the less, Chapters 5 and 6 give an excellent guideline on hand how to hydrolyze proteins enzymatically in a classical (but strictly controlled) way (Chapter 5) and in an online format with subsequent mass spectrometric detection (Chapter 6).

Seronei Chelulei Cheison (by the way with Kenyan roots) and Ulrich M. Kulozik describe the professional protein digest with proteases giving detailed information about systematic regulation and properties which are all too often ignored in 'overnight digestions'. Johannes Hoos and Wilfried M.A. Niessen introduce a well-structured and novel technique for online-digest with subsequent MS-detection of proteolytic peptides.

In Chapter 7 the colleagues Christian Webhofer and Michael Schrader give a very detailed insight in using state-of-the art bioinformatic software tools. As an example they chose the quantitative strategy with isotope labeling and its bioinformatic handling.

Chapter 8 is written by Gabriele Stöhr and Andreas Tebbe, who obviously know what a hard job it is to obtain quantitative data in protein analysis. They describe a SILAC experiment with all the needs and problems to obtain high quality data. Their clear presentation allows the reader to form their own opinion about this type of proteomics.

Boris Macek in Chapter 9 presents various novel strategies for the extraction and the detection of protein phosphorylation. His experience and its fast conversion can be seen in his impressively fast delivering of this chapter to the Editor! – To be honest though, I am very thankful to all the authors for delivering their chapters fast and in time as well as for the high quality of each single chapter.

Chapters 10 and 11 represent a new view on applying LC-MS on proteins. Both chapters deal with the observation of protein function (*e.g.* enzymatic function in Chapter 10 and non-covalent interaction in Chapter 11).

The local colleagues Romy K. Scheerle and Johanna Graßmann provide in Chapter 10 very detailed information about the novel observation of enzymatic activity and regulation by microflow and nanoflow coupled mass spectrometry. They give a crucial guideline for a future key technology in perspectives of 'functional proteomics'.

In Chapter 11 Michael Krappmann and I present further continuous flow mixing systems for the online screening of enzymatic activity regulation and – as shown therein – for the detection of non-covalent protein complexes. The chapter furthermore introduces the first application of the specially developed analysis software Achroma (freely available from the authors!) to process the mass spectrometric untypically raw data.

Last but not least, Chapter 12 is written by my long-term colleague Rene Wissiack. For many years he has worked in the Austrian industry and knows all about the commonly-secret problems with industrial protein analysis problems. Thus I am really happy, that he wrote a chapter about the industrial needs, although he does not show confidential examples of pragmatic industrial solutions. In order to avoid any possible conflicts of interest there are no Notes and Perspectives formulated in this chapter.

In general, my philosophy to this book was to restrict the authors as little as possible regarding style and outline, because everybody has his or her own style to teach their experimental knowledge. Thus I am sure that the small variations between the various chapters do not confuse, but rather help to transfer the knowledge most effectively. Furthermore this feature may help to find an own preferred teaching style for each reader and also a favored style of knowledge transfer.

This directly leads me to an offer to you as a reader: I invite you very cordially to write a practical and applicable chapter about your expertise in LC-MS of proteins and peptides. If you send us this chapter we will be able to publish an extended version of this book with further experimental strategies. Independently, I hope with the current edition to enable a fast and unproblematic start into the proteomics sector for many new 'players'. Furthermore we are optimistic that also experienced proteomics people can extend their practical knowledge with some new hints and tips.

Finally, I want to thank a lot of people for their help and input. Again, I thank all the authors for their really straightforward work and their impressive professional contributed knowledge. I thank Merlin Fox from the RSC for his long-term interest in this practical book and the opportunity to create it in the current flexible but 'red-lined' style. Rosalind Searle from the RSC is thanked for her constructive and relaxed input.

Last, but not least, I thank Ingolf Krause, my first professional teacher in protein purification, for showing me how to handle proteins – softly but strictly! In memorandum to this excellent protein analyst I want to dedicate this book to him.

Thomas Letzel
Freising, Germany

Contents

RSC Chromatography Monographs No. 15
Protein and Peptide Analysis by LC-MS: Experimental Strategies
Edited by Thomas Letzel
© The Royal Society of Chemistry 2011
Published by the Royal Society of Chemistry, www.rsc.org

Contributors

Seronei C. Cheison Zentralinstitut für Ernährungs- und Lebensmittel-forschung (ZIEL): Bioactive Peptides and Protein Technology, Technische Universität München, Weihenstephaner Berg 1, D-85354 Freising, Germany and School of Public Health and Community Development, Maseno University, Private Bag, Kisumu, Kenya

Tamo Fukamizo Kinki University, Department of Advanced Bioscience, 3327–204 Nakamachi, Nara, 631–8505, Japan

Benjamin A. Garcia Princeton University, Department of Chemistry and Department of Molecular Biology, Princeton, NJ 08544, USA

Johanna Graßmann Institute for Chemical-Technical Analysis and Chemical Food Technology, Center of Life and Food Sciences Weihenstephan, Technische Universität München, Weihenstephaner Steig 23, 85350 Freising, Germany

S. Johannes Hoos VU University, Faculty of Sciences, BioMolecular Analysis Group, De Boelelaan 1083, 1081 HV, Amsterdam, The Netherlands

Michael Krappmann Institut für Forschung und Weiterbildung, University of Applied Sciences, Hochschule Weihenstephan – Triesdorf, Am Hofgarten 4, 85350 Freising, Germany

Ulrich M. Kulozik Zentralinstitut für Ernährungs- und Lebensmittel-forschung (ZIEL) Abteilung Technologie, Lehrstul für Lebensmittelverfah-renstechnik und Molkereitechnologie, Technische Universität München, Weihenstephaner Berg 1, D-85354 Freising, Germany

Thomas Letzel Competence Pool Weihenstephan (CPW), Center of Life and Food Sciences Weihenstephan, Technische Universität München, Weihenstephaner Steig 23, 85350 Freising, Germany

Friedrich Lottspeich Max Planck Institute of Biochemistry, Protein Analysis, Am Klopferspitz 18, 82152 Martinsried, Germany

RSC Chromatography Monographs No. 15
Protein and Peptide Analysis by LC-MS: Experimental Strategies
Edited by Thomas Letzel

Published by the Royal Society of Chemistry, www.rsc.org

Boris Macek Proteome Center Tübingen, Interdepartmental Institute for Cell Biology, University of Tübingen, Auf der Morgenstelle 15, 72076 Tübingen, Germany

Wilfried M.A. Niessen VU University, Faculty of Sciences, BioMolecular Analysis Group, De Boelelaan 1083, 1081 HV, Amsterdam, The Netherlands

Takayuki Ohnuma Kinki University, Department of Advanced Bioscience, 3327–204 Nakamachi, Nara, 631–8505, Japan

Romy K. Scheerle Institute for Chemical-Technical Analysis and Chemical Food Technology, Center of Life and Food Sciences Weihenstephan, Technische Universität München, Weihenstephaner Steig 23, 85350 Freising, Germany

Michael Schrader Weihenstephan-Triesdorf University of Applied Sciences, Department of Biotechnology and Bioinformatics, 85350 Freising, Germany

Gabriele Stöhr Max Planck Institute of Biochemistry, Department of Proteomics and Signal Transduction, Am Klopferspitz 18, 82152 Martinsried, Germany

Andreas Tebbe KINAXO Biotechnologies, Am Klopferspitz 19a, 82152 Martinsried, Germany

Christian Webhofer Max Planck Institute of Psychiatry, Proteomics and Biomarkers, Professor Dr. Christoph W. Turck, Kraepelinstrasse 2-10, 80804 München, Germany

Rene Wissiack Biotech Operations, Process Science, In-Process-Control, Boehringer Ingelheim RCV GmbH & Co KG, Dr. Boehringer-Gasse 5-11, 1121 Wien, Austria

Nicolas L. Young Princeton University, Department of Molecular Biology, Princeton, NJ 08544, USA

CHAPTER 1

Top Down and Bottom Up Analysis of Proteins (Focusing on Quantitative Aspects)

FRIEDRICH LOTTSPEICH

Max Planck Institute of Biochemistry, Protein Analysis, Am Klopferspitz 18, 82152 Martinsried, Germany

1.1 Introduction

One key focal point in proteome research is the determination of changes in protein expression and their modifications. In the early years of proteomics the field was dominated by protein chemists and the main approach was 2D-PAGE where differential maps revealed protein pattern differences. The detailed analysis of the different protein spots only became feasible after the introduction of mass spectrometry. However, 2D-PAGE was difficult to reproduce, was not automated and had several limitations with important subsets of proteins (*e.g.* hydrophobic, very basic, very large or very small proteins). Furthermore, the quantification of the proteins was usually performed by image analysis following several staining methods, which exhibit different signal intensities with different proteins. The dynamic range of detection spans only about 2–3 orders of magnitude, resulting in the visualization of only relatively highly abundant proteins. Additionally, image analysis in principle cannot deal with protein mixtures in a single spot, which, due to the complexity of a proteome, is the common case. Finally, enzymatic cleavage of the protein in the gel matrix suffered from low peptide recovery.

RSC Chromatography Monographs No. 15
Protein and Peptide Analysis by LC-MS: Experimental Strategies
Edited by Thomas Letzel
© The Royal Society of Chemistry 2011
Published by the Royal Society of Chemistry, www.rsc.org

All these limitations encouraged mass spectrometric experts to develop alternative strategies for proteome analyses. Mass spectrometry was used to work with small molecules and therefore it was tempting to cleave the very heterogeneous and unpleasant protein complexity of a proteome enzymatically into small peptides (see also Chapter 5) with much more favourable properties concerning hydrophobicity, diversity and accessibility for multidimensional chromatographic separations and mass spectrometry (see also Chapter 6). Soon mass spectrometry together with informatics and protein databases were developed and optimized to handle complex peptide mixtures, allowing peptide identification by high-throughput MS-MS (see also Chapter 7). The mainstream of proteome research followed this track, called 'bottom up' or 'shotgun' proteomics (Figure 1.1).

Unfortunately, the bottom up approach also has several severe and fundamental limitations. First, by cleaving the proteome into peptides the complexity increases by a factor of about 40, producing hundreds of thousands of peptides. This is a number that swamps even the most modern mass spectrometers. In consequence, only a fraction of these peptides can be analysed in detail ('undersampling' effect), and it is difficult to assure that identical peptides are analysed from each sample, which is essential to unravel a quantitative fluctuation in the amounts of certain proteins. Second, and even more serious, the context of a protein and the derived peptides is destroyed. A certain peptide may be derived from different proteins or from different forms of a certain gene product, such as post-translational modified, processed or truncated protein species, or from

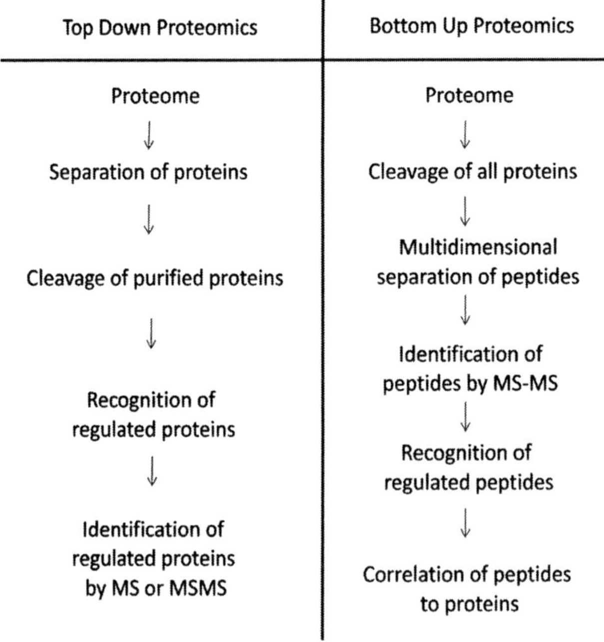

Figure 1.1   'Top down' *versus* 'bottom up' proteomics strategies.

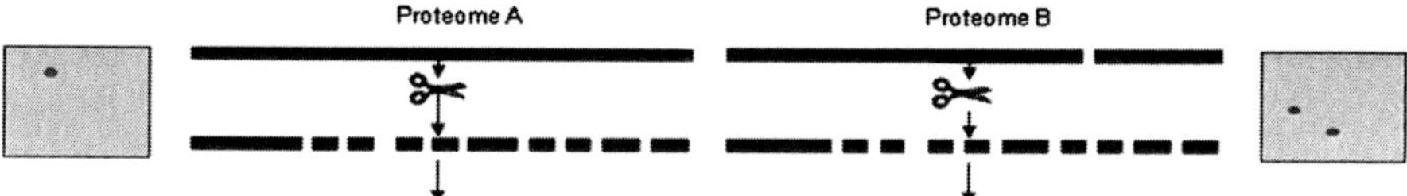

Figure 1.2 Top down *versus* bottom up approach of a certain protein processed in proteomic state B. Applying a top down approach (*e.g.* a 2D gel) the cleavage event is easily seen. In a peptide-based approach the two situations (A and B) give identical peptide patterns.

proteins having in common major amino acid sequence stretches like splicing variants or protein isoforms. A single gene will almost always produce an unpredictable multiplicity (tens or even hundreds) of different protein species, which are composed predominantly of identical peptides. Consequently, the quantitative analysis of a peptide monitors only the sum of all proteins that contain this particular peptide. Unfortunately, usually it is not known which protein species are expressed at a certain proteome state. Since in proteome analysis the sequence coverage usually is far less than 50%, the probability of missing the nature of the diversity or the modifications is rather high. In conclusion, the quantity of a peptide determined by a bottom up approach does not necessarily reflect the quantity of a protein of interest. This is completely different in a protein-based, *i.e.* 'top down', approach (see also Chapter 4). The molecular structure and the nature of an intact protein are well defined by molecular properties like the molecular mass and the position in a separation space (isoelectric point, chromatographic position, *etc.*). If, for example, one compares two proteome states, where a single protein is processed by a cleavage after a lysine, this is easily seen on a protein-based (top down) analysis, like 2D-PAGE or even 1D-PAGE. However, with a bottom up approach, in the two situations all the peptides appear exactly the same and the biological difference will not be detectable (Figure 1.2).

Despite these obvious limitations, bottom up approaches are widely used. The relative technical simplicity and the enormous instrumental development tailor-made towards peptide-based approaches, on the separation as well as on the mass spectrometry side, forced the proteomics field to go mainly with bottom up strategies. However, in recent years the importance of the protein diversity caused by post-translational modifications (PTM; see also Chapter 9), degradations and processing events has become evident to proteome scientists. Still, major technical hurdles have to be vanquished but the awareness of the potential and the advantages of top down proteomic approaches is significantly increasing.

1.2 Quantitative Proteomics

1.2.1 Quantitative Proteomics by Label-Free Techniques

Mass spectrometry *per se* is not an absolutely quantitative technique. Sequence-dependent peptide ionization efficiencies and suppression of neighbouring signals by dominant peptides results in a low correlation between peptide mass signal intensity and the amount of the peptide (see also Chapter 2). Especially with highly complex mixtures, as commonly achieved in proteomes studies, this inhibits an easy and direct quantification of peptides by signal intensity. However, recently label-free LC-MS quantification methods have been described to determine relative abundances of proteins between multiple conditions. 'Spectral counting' methods based on the number of spectra for a certain protein found in a proteome analysis does correlate quite well with the protein amount and thus may provide an estimation on the relative protein amount.[1] With highly abundant proteins the response in spectral increasing protein amount is saturable. With proteins of low abundance the data at low spectral count are noisy and the sensitivity for fold changes decreases.

More accurate is the modification of the spectral count approach by Silva *et al.*[2] The authors found that a protein's abundance could be well estimated from the average mass spectrum peak intensity of its three best-detected peptides, assuming that the signal intensity of a fully ionized peptide is roughly dependent on the protein amount.[2]

All label-free proteomics approaches published so far are based on bottom up proteomics strategies and are successful on relatively non-complex samples with almost no sample preparation applied. However, in a more complex situation like human tissues or body fluids multidimensional sample preparation is the key issue to reduce the complexity to a tolerable level. With label-free proteomics a separate analysis has to be performed for each proteome state and no multiplexing is possible; the whole separation and analysis workflow has to be performed for each sample individually. We may expect it to be extremely difficult to reach the required quantitative reproducibility, at least with top down proteomics approaches.

1.2.2 Quantitative Proteomics by Isotopic Labelling Techniques

For many years, metabolic studies have used non-radioactive isotope labelling combined with mass spectrometry as a powerful tool for quantification. Analogues of the metabolites to be tested were synthesized containing ^{13}C, ^{15}N or ^{2}H, and were spiked in defined amounts into the biological sample. As isotopic variants of all molecules behave identically and exhibit the same ionization behaviour during an experiment, quantification by signal intensity of isotopologues is highly accurate. This successful concept was transferred to proteomics. Two or more protein or peptide samples are differentially labelled, one with an isotopically 'light' and the others with isotopically 'heavy' tags. The samples are then combined, thereby 'freezing' the relative amount of proteins or peptides. The complexity of the samples is then reduced by using one or more separation steps. After reduction of complexity and enzymatic cleavage, peptides resulting from corresponding proteins of both samples retain the same chemical properties despite being differentially labelled. A certain peptide from different proteome states can be detected as a mass pair or a mass multiplet

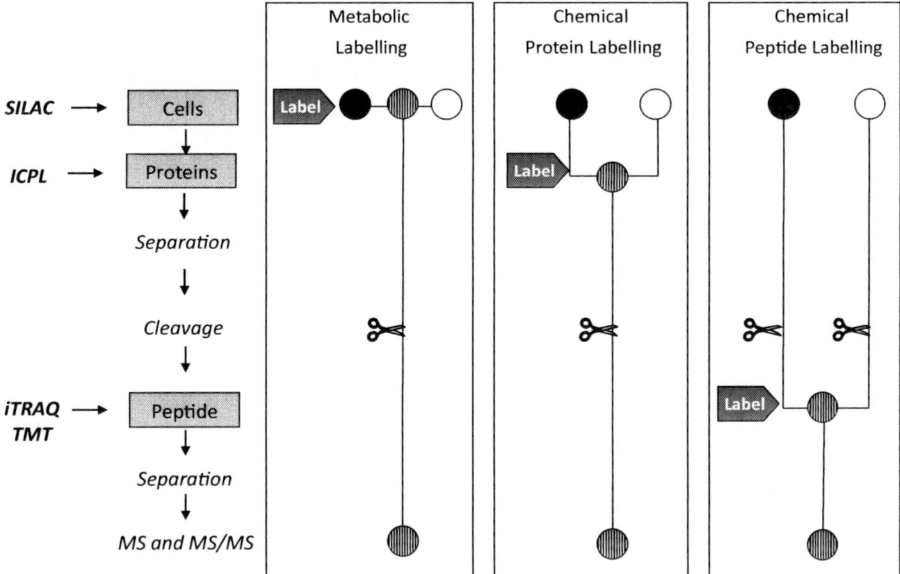

Figure 1.3 Labelling strategies for stable isotope incorporation. The workflows differ in the time point when stable isotopes are introduced into proteins or peptides. Sample (●, heavy isotope) and control (○, light isotope) are labelled separately and combined afterwards.

(two or more proteome samples) during mass spectrometry, differing only by the masses introduced by the isotopic labels. Corresponding peak heights or areas are then compared to calculate the relative abundance of corresponding peptides of the different samples.

The crucial difference between different labelling strategies is the time point of incorporation of the isotopic label. Prior to the labelling step, samples have to be processed independently in parallel. However, any reaction or handling step not performed under isotopic control may result in quantification errors.

The approaches most often at present are summarized in Figure 1.3. Subsequently their advantages and limitations will be discussed briefly. Additionally, an exemplary isotopic labelling experiment can be found in Chapter 7 with a detailed method description.

1.2.2.1   Introduction of the Isotopic Label at the Level of Living Cells (Metabolic Labelling)

The main advantage of the metabolic labelling strategy is that the label is introduced into living cells by *in vivo* incorporation of amino acids containing stable isotopes. Therefore, cells from different states, following differential labelling, can be mixed before lysis. Subsequent steps of fractionation and purification do not affect the accuracy of quantification. Consequently, stable isotope labelling by amino acids in cell culture (SILAC) has become one of the most widely used strategies in quantitative proteomics.[3] Two or more cell populations are simply grown in different media, each containing a light version or one or more heavy versions of a suitable amino acid. Several amino acids are described as being used in the SILAC approach. Labelling of arginine and lysine, followed by tryptic digestion, results in labelling of almost every peptide except the C-terminal of each protein. The use of other amino acids such as tyrosine or methionine has also been described. SILAC is mainly used for cell culture-based proteomics approaches. An example of a SILAC experiment with detailed descriptions can be found in Chapter 8.

The *in vivo* incorporation of stable isotopes has been demonstrated even in animals.[4] One significant limitation of SILAC is that it cannot easily be used for samples which are not grown in culture. Samples obtained from patients (*e.g.* tissues) can only be analysed by the addition of an artificial internal standard composed of a relevant mixture of cell lines that somehow can resemble the protein content of the actual tissue. Samples from body fluids can only be routinely quantitatively analysed by use of chemically introduced isotopic tags.

1.2.2.2 Introduction of Chemical Tags to Complex Protein Mixtures

When using a chemical-based labelling approach, stable isotope-bearing reagents react with the reactive sites (SH- or amino groups) of a protein. In 1999, Gygi and colleagues introduced this new approach based on chemical labelling using isotope-coded affinity tags (ICAT) directed to cysteine residues.[5] However, cysteine is a rather rare amino acid. Therefore, after enzymatic cleavage only a few peptides carry the isotopic label, *i.e.* the quantitative information. Thus, with this technique the sequence coverage remained marginal and it is very little used now. A more extended, robust and complete labelling was obtained with the amino group directed ICPL label, which has become the predominant reagent for isotope-labelled top down proteomic approaches, especially since dedicated software (ICPL*Quant*) has been developed covering the whole workflow and the automated quantitative data analysis.[6,7] The ICPL method allows for up to fourfold multiplexing and provides highly accurate and reproducible quantification, high protein sequence coverage, including PTMs and isoforms, and is compatible with all commonly used protein and peptide separation techniques. Two or more protein mixtures obtained from different proteomic states are individually reduced and alkylated to denature the proteins and to ensure easier access to free amino groups which are subsequently derivatized with the ^{12}C (light), ^{2}H (medium), ^{13}C (heavy) and ^{13}C^{2}H (ultraheavy) variants of the ICPL reagent. After combining the mixtures, any separation method can be applied to reduce the complexity of the sample on the protein level. Isoelectric focusing by OffGEL or 2D gel electrophoresis may be used as high-resolution separation technologies, where especially protein isoforms can be well distinguished. After significant reduction of complexity the protein fractions are enzymatically digested, preferably using a double enzyme approach.[8] The resulting peptides are quantified by mass spectrometry. Identical peptides derived from the differently labelled protein samples differ in mass and thus can be assigned to the corresponding proteomic state. Each lysine-containing peptide will appear as a multiplet in the acquired MS spectra. The ratios of the ion intensities of these sister peptide multiplets allow for the determination of the relative abundance of their parent proteins in the original samples. After relative quantification only differently regulated proteins have to be identified either by peptide mass fingerprint (PMF) or CID.

1.2.2.3 Introduction of Chemical Tags in Complex Peptide Mixtures

Isotopic labelling methods have also become popular for bottom up strategies, to achieve a more accurate quantification.

Amino group directed non-isobaric reagents such as ICPL are well suited for this purpose.[9] However, the concept of these reagents is based on the relative quantification of stable isotope-labelled peptides prior to MS-MS analysis. Since with bottom up proteomics approaches the complexity is already increased significantly by the enzymatic cleavage of the proteome, a further increase in complexity is caused by the different isotopic derivatives of the proteomics states. As a consequence, many peptides coelute during chromatography of complex biological samples, causing signal suppression in mass spectrometry and making quantitative interpretation and identification difficult. To avoid a further increase in peptide complexity and at the same time allow for higher multiplexing, an approach using isobaric isotope reagents was recently introduced.

1.2.3 Isobaric Tags for Relative and Absolute Quantification

The core of this methodology is a multiplexed set of up to eight isobaric reagents (iTRAQ, AB-Sciex; TMT, ThermoScientific).[10,11] The labels consist of an *N*-hydroxysuccinimide moiety, reacting with amino groups, and two isotope-coded regions, a balance group and a reporter group. The latter is released during MS-MS, yielding MS signals at 113–121 Da (ITRAQ) or 126–131 Da (TMT). Corresponding isobaric peptides of all the proteome states coelute during chromatography and are indistinguishable in MS, but exhibit low-mass MS-MS signature ions (reporter ions) that support relative peptide quantification. Since for quantification, MS-MS spectra of each peptide are needed, tens of thousands of MS-MS spectra per analysis are necessary. Furthermore, for a correct quantification no coeluting isobaric peptides should be present, which is hard to achieve without reduction of complexity. Therefore, these approaches show promise mainly with proteomic samples of rather low complexity.

1.2.4 Absolute Quantification in Proteomics with Targeted Proteomics

Most proteomics projects so far have been performed using relative quantification, *i.e.* only monitoring changes in the level of a large number of proteins.

However, for a deeper understanding of biological situations and modelling purposes, as in systems biology, it is often necessary to know the absolute amount of certain proteins. This is in most cases only possible with targeted proteomics techniques. Unfortunately, an internal labelled reference protein standard for the protein(s) of interest is usually not available. To circumvent this, peptides contained in a protein may serve as surrogate markers for the protein itself. Peptides can easily be synthesized with isotopic amino acids or can be reacted with isotopic reagents to serve as internal standards. Therefore, today targeted proteomics approaches are typical bottom up proteomics approaches with all the limitations described above. Thus, multiplexed isotopic labelled peptide-based approaches offer the possibility of performing absolute quantification by using one label for the peptide mixture of defined amounts of synthetic peptides contained in the proteins of interest. Several elegant methods have been proposed for the cost-effective and accurate production of the standard sample.[12,13]

1.2.5 Absolute Quantification Using SRM

The increasing demand for absolute quantification and monitoring certain proteins was also addressed by instrument manufacturers developing dedicated instruments and workflows for selected reaction monitoring (SRM).[14,15]

In SRM assays the first (Q1) and last (Q3) mass analysers of a triple quadrupole mass spectrometer are used as mass filters to isolate a peptide ion and a corresponding fragment ion, respectively. The signal of the fragment ion is then monitored. The selectivity achieved by using two filtering stages results in a quantitative analyses with unmatched sensitivity down to the (sub)attomole level. High specificity can be obtained by monitoring the fragment ion signals only over the expected chromatographic elution time of the peptide. The specific pairs of m/z values associated to the precursor and fragment ions selected are referred to as 'transitions' and effectively constitute mass spectrometric assays that allow the identification and quantification of a specific peptide. However, for reliable peptide identifications several transitions of a single peptide have to be measured. To obtain quantitative protein results usually three or more peptides belonging to the same protein have to be monitored in a complex protein digest. An SRM-like evaluation may also be performed using a new generation of mass spectrometers acquiring high-quality LC-MS-MS spectra of a complex peptide mixture. High-resolution (*e.g.* 30 000) quantitative data at high mass accuracy which are acquired at high speed (*i.e.* 50 MSMS spectra/s) allow postanalytical data mining to filter any desired MS-MS transition for peptide quantification. SRM assays are powerful targeted proteomics techniques, especially since the assay development is significantly facilitated by comprehensive databases for transitions, such as Peptide Atlas, which already cover several transitions of almost all yeast and human proteins.[16,17] SRM is believed to be widely used in biology and clinics and is predicted to become a serious competitor for immunological assays.

1.3 Notes

At present, for general quantitative proteomics approaches SILAC represents the method of choice when the *in vivo* introduction of isotopologues can be achieved. For samples of body fluids and tissues, or for any pathological sample, the ICPL technology provides highly accurate results. Both top down proteomics approaches have the capability to recognize and quantify splice variants, isoforms and posttranslational modified protein species. This is certainly a fundamental advantage compared to peptide-based bottom up proteomics approaches, where all these protein species can hardly be reliably quantified. However, shotgun proteome analyses using isobaric labels, like iTRAQ or TMT, allow for higher multiplexing and monitoring of a large number of samples, dominating the 'bottom up' field. These techniques are in strong competition with label-free techniques, which will have to prove their suitability for complex protein mixtures in future.

Probably the most advanced and promising techniques in proteomics are targeted approaches like SRM, which even allow the absolute quantification that is urgently needed especially in the emerging field of systems biology. With these targeted techniques and with major efforts of the scientific community to systematically provide peptide mass data, like ProteinAtlas, and developing innovative mass spectrometric analysis strategies, it will soon become feasible to monitor complete and complex networks and provide valuable data for systems biology modelling.

References

1. K. Kito and T. Ito, *Curr. Genomics*, 2008, **9**, 263.
2. J. C. Silva, M. V. Gorenstein, G. Z. Li, J. P. C. Vissers and S. J. Geromanos, *Mol. Cell. Proteomics*, 2006, **5**, 144.
3. S. E. Ong, B. Blagoev, I. Kratchmarova, D. B. Kristensen, H. Steen, A. Pandey and M. Mann, *Mol. Cell. Proteomics*, 2002, **1**, 376.
4. C. C. Wu, J. M. MacCoss, K. E. Howell, D. E. Matthews and J. R. Yates, *Anal. Chem.*, 2004, **76**, 4951.
5. S. P. Gygi, B. Rist, S. A. Gerber, F. Turecek, M. H. Gelb and R. Aebersold, *Nat. Biotechnol.*, 1999, **17**, 994.
6. A. Schmidt, J. Kellermann and F. Lottspeich, *Proteomics*, 2005, **5**, 4.
7. A. Brunner, E. Keidel, D. Dosch, J. Kellermann and F. Lottspeich, *Proteomics*, 2010, **10**, 315.
8. A. Turtoi, G. D. Mazzucchelli and E. De Pauw, *Talanta*, 2010, **80**, 1487.
9. M. Fleron, Y. Greffe, D. Musmecii, A. C. Massart, V. Hennequiere, G. Mazzucchelli, D. Waltregny, M. C. De Pauw-Gillet, V. Castronovo, E. De Pauw and A. Turtoi, *J. Proteomics*, 2010, **73**, 1986.
10. P. L. Ross, Y. N. Huang, J. N. Marchese, B. Williamson, K. Parker, S. Hattan, N. Khainovski, S. Pillai, S. Dey, S. Daniels, S. Purkayastha, P. Juhasz, S. Martin, M. Bartlet-Jones, F. He, A. Jacobson and D. J. Pappin, *Mol. Cell. Proteomics*, 2004, **3**, 1154.

11. A. Thompson, J. Schafer, K. Kuhn, S. Kienle, J. Schwarz, G. Schmidt, T. Neumann and C. Hamon, *Anal. Chem.*, 2003, **75**, 1895.
12. O. Stemmann, H. Zou, S. A. Gerber, S. P. Gygi and M. W. Kirschner, *Cell*, 2001, **107**, 715.
13. J. Holzmann, P. Pichler, M. Madalinski, R. Kurzbauer and K. Mechtler, *Anal. Chem.*, 2009, **81**, 10254.
14. N. R. Kitteringham, R. E. Jenkins, C. S. Lane, V. L. Elliott and B. K. Park, *J. Chromatogr., B*, 2009, **877**, 1229.
15. V. Lange, P. Picotti, B. Domon and R. Aebersold, *Mol. Syst. Biol.*, 2008, **4**, 1.
16. Peptide Atlas, www.peptideatlas.org.
17. E. W. Deutsch, J. K. Eng, H. Zhang, N. L. King, A. I. Nesvizhskii, B. Y. Lin, H. K. Lee, E. C. Yi, R. Ossola and R. Aebersold, *Proteomics*, 2005, **5**, 3497.

How to Couple and Handle Liquid Chromatography with Mass Spectrometry

THOMAS LETZEL

Competence Pool Weihenstephan (CPW), Center of Life and Food Sciences Weihenstephan, Technische Universität München, Weihenstephaner Steig 23, 85350 Freising, Germany

2.1 Introduction

2.1.1 Separation

Today, various chromatographic separation techniques can be coupled with MS: gas chromatography (GC), high performance liquid chromatographic techniques (HPLC) like reversed-phase liquid chromatography (RP-LC), hydrophilic interaction liquid chromatography (HILIC), hydrophobic interaction chromatography (HIC), and capillary electrophoresis (CE) with its subclasses. Less suitable for coupling are gel filtration chromatography (GFC) also known as size-exclusion chromatography (SEC), immune affinity chromatography (IAC), ion-exchange chromatography (IEC), and isoelectric point chromatography (IPC). However, most of them are used, especially in protein purification and peptide separation. The experimental handling of IEC, GFC and HIC for protein purification is described in more detail in Chapter 3 and the handling of RP-LC, IEC, SEC and HILIC in Chapter 4.

RSC Chromatography Monographs No. 15
Protein and Peptide Analysis by LC-MS: Experimental Strategies
Edited by Thomas Letzel
© The Royal Society of Chemistry 2011
Published by the Royal Society of Chemistry, www.rsc.org

GC can easily be coupled with MS, because both are gas-phase techniques. Because they use the same gaseous environment there is no problematic interaction, whereas the success of liquid phase separation coupled to MS depends on the transfer of dissolved analytes to the gas phase. The hyphenation is often of low efficiency due to the incompatibility of many liquid mobile phases with mass spectrometric conditions. On the other hand, great efforts have been made to ensure a sensitive and highly efficient transfer of analytes because HPLC techniques are highly capable of separating proteins and peptides. Additionally to the RP-HPLC described in this chapter, the separation techniques described above are partially compatible with mass spectrometric detection.

2.1.2 Ionization

Effective transfer of charged molecules *via* 'ion sources' into the mass spectrometer was (and still is) the key process for sufficient investigation of molecules from different origins after either a chromatographic separation or from surfaces (Figure 2.1).

For example, the ionization techniques of electron impact (*vs* electron ionization, EI) and chemical ionization (CI) are well-established for GC-separated analytes. However, this use of ionization for biomolecules is restrictedly by the

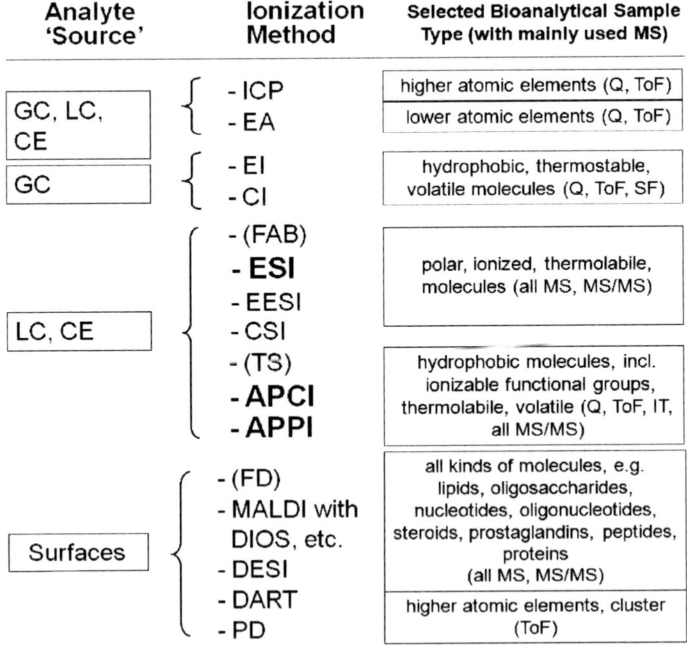

Figure 2.1 Types of ionization sources with connected separation techniques and selected examples in biological applications.

harsh ionization conditions. On the other hand, the same ionization principle was originally applied for analytes eluting from LC, so low-pressure ionization chambers like thermospray and fast-atom bombardment (FAB) were used. Again, the ionization sources were only partly useful for (large and polar) biomolecules.

Since new-generation atmospheric pressure ionization (API) sources became commercially available in the 1990s, the liquid-dissolved analytes can be transferred at higher ionization rates. These 'ion sources' are typically electrospray ionization (ESI), atmospheric pressure chemical ionization (APCI), and atmospheric pressure photo-ionization (APPI). Figure 2.2 illustrates ESI (conventional (a) and newly developed (b)), APCI (c), APPI (d) and multimode ionization (e; MMI, *i.e.* a combination of ESI and APCI) sources. ESI is of special interest for the detection of intact proteins, peptides, carbohydrates and other large molecules due to the soft transfer of charged molecules from solution into the gas phase. In contrast APCI and APPI are of increasing importance because of their advantages in detecting either small organic (and originally uncharged) molecules or hydrophobic but stable peptides and others.

Recent developments in the elemental speciation and quantification have led to the introduction of new LC-MS systems. In this context inductively coupled plasma ionization (ICP) is a very well established technique for the characterization of biomolecules containing higher atomic elements. Further on, the elemental analyser (EA) can be applied for the detection of lower atomic elements and the rarely used plasma desorption (PD) allows the ionization of cluster containing several elements. The latter system is a so-called 'non-chromatographic coupled' ionization technique that transfers ions from a fixed surface into the gas phase. Further ionization techniques in this field are the 'grandmother' field desorption (FD), the frequently used matrix-assisted laser desorption/ionization (MALDI), with its branches desorption/ionization on silicon (DIOS), self-assembled monolayers for desorption/ionization (SAMDI) and surface-enhanced laser desorption/ionization (SELDI), desorptive electrospray (DESI), direct analysis real time (DART) and others.

ESI, APCI, APPI, EESI (extractive electrospray ionization) and CSI (cold spray ionization) are currently the techniques typically coupled to MS detection if the analytes are diluted in liquids. Among these ESI[1] is the most frequently used, whereas APCI[2] and APPI[3] as well as EESI[4] and CSI[5] are used for more specialized molecule applications.

2.1.3 Mass Spectrometric Detection

Typical standard mass spectrometers contain the single mass analyser quadrupole (Q), time-of-flight (ToF) or sector field (SF). They are well known and robust systems due to their long history as GC-coupled MS systems. Also, the hyphenation of liquid-phase techniques with Q and ToF mass spectrometers is well known and nowadays mostly used for thermolabile and polar molecules.

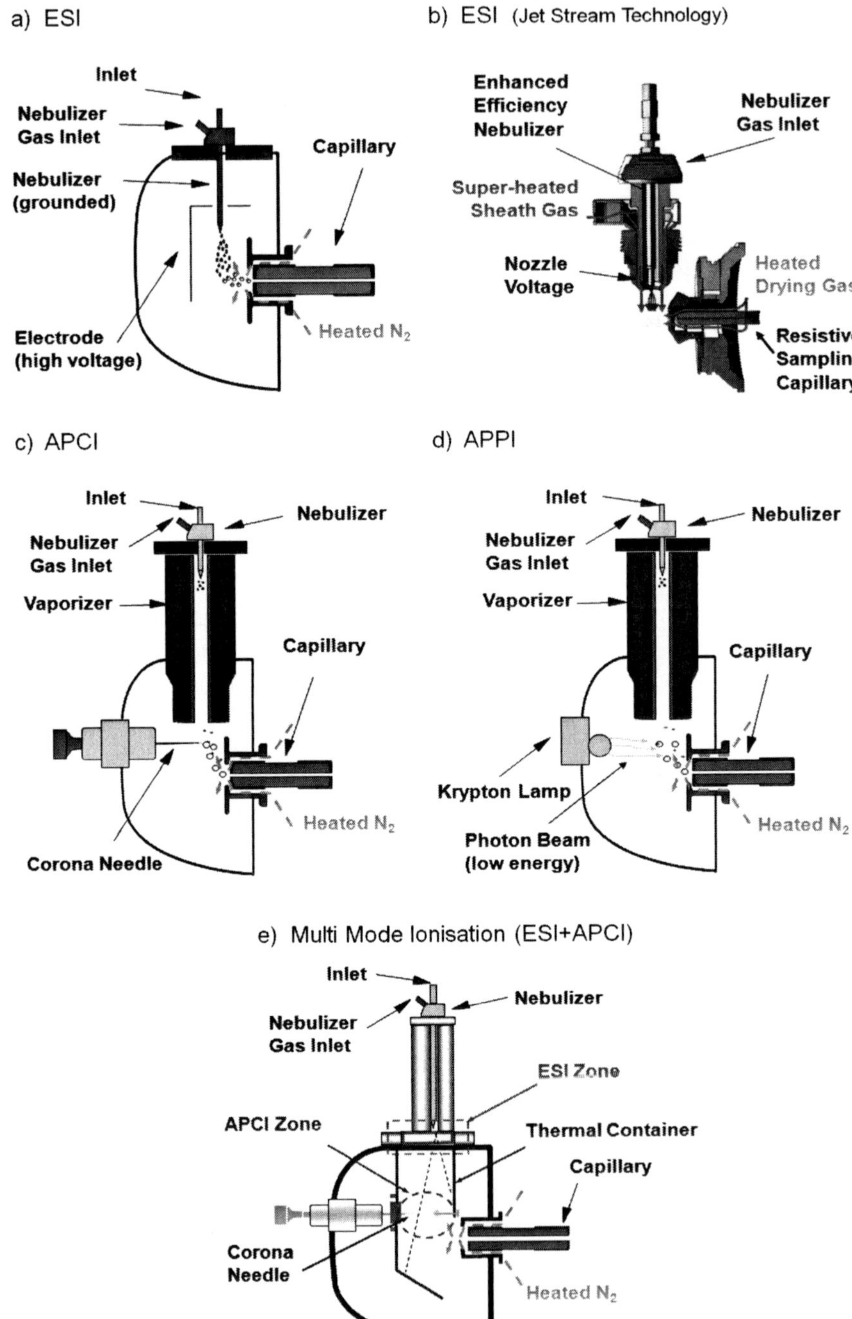

Figure 2.2   Layouts of ESI: (a) conventional, (b) newly developed, (c) APCI, (d) APPI (e) multimode ionization source (MMI, *i.e.* a combination of ESI and APCI). Reprinted after modification with permission of Agilent Technologies Inc. Copyright 2010, Agilent Technologies, Inc.

A huge advancement was the combination of those analysers with so-called tandem mass spectrometers ('tandem in space'). Tandem mass spectrometers have ion extraction regions, fragmentation regions and fragment separation regions with subsequent ion detection. The ion extraction and the fragment separation are mainly performed with the mass analysers Q or ToF, whereas for a long period time the fragmentation region contained exclusively a Q. Consequently, QqToF and QqQ- and later ToF-ToF setups are utilized for identification and/or sensitive quantification. Electrospray ionization interfaces are often coupled with QqToF, whereas ToF-ToF mass spectrometers are mainly sold with MALDI interfaces because of their perfect compatibility.

For a long time ion trap analyzers (IT, a 'tandem in time' analyser) were the only type of analyser that could detect analyte fragments in higher order MS (MS^n). However, the popularity of this type of analyser is increasing again, since ITs are on the market as linear ion trap (LIT), Fourier transform-ion cyclotron resonance (FT-ICR) and orbitrap mass spectrometers. The last two possess very high resolution, extreme accuracy and very low detection limits. Thus, a 14.5 T ESI-Q FT-ICR mass spectrometer is currently the most powerful system available with a mass resolution higher than 2×10^6, an accuracy <1 ppm, sensitivities in the attomole region (and, by the way, the highest price).

Originally the fragmentation region (a Q) in tandem mass spectrometers was utilized in its function as a collisional-induced dissociation cell (CID; also known as CAD, collisional-activated dissociation). However, in recent years the variety of fragmentation techniques for tandem mass spectrometers (in space) has become immensely extended. The CID is now in strong competition with fragmentation methods having different 'specificities', like blackbody infrared radiative dissociation (BIRD), electron-capture dissociation (ECD), electron transfer dissociation (ETD), infrared multiphoton dissociation (IRMPD) and modified ion mobility cells.

Further information can be found in excellent books[6,7] and articles[8] on MS.

2.2 Materials and System

2.2.1 Chemicals

- The oxy-PAH 9,10-anthraquinone was purchased from Merck-Schuchardt (Hohenbrunn/München, Germany)
- The protein malantide was obtained from Sigma-Aldrich (Steinheim, Germany)
- LC-MS grade methanol and water were obtained from J.T. Baker (Deveter, Holland)

2.2.2 HPLC and Mass Spectrometer

- For the experiments, a HPLC system (1200 series, Agilent Technologies, Waldbronn, Germany) was used in combination with a triple quadrupole

mass spectrometer (6410, Agilent Technologies, Waldbronn, Germany). HPLC, ionization, and the MS detector were controlled and data was analysed by MassHunter Workstation software (version B.02.01, Qualitative Analysis (version B.02.00, Agilent Technologies, Waldbronn, Germany)
- Special test solutions were used for the application of ESI and APCI:
  - Solution 1 (ESI positive/negative): 7.5 µM; 15 µM; 30 µM malantide in methanol
  - Solution 2 (APCI positive/negative): 50 µM; 75 µM; 100 µM 9,10-anthraquinone in methanol

2.2.1.1   *Experimental HPLC Condition and MS Parameter*

- HPLC conditions:
  - HPLC: Agilent Technologies, 1200 Series
  - HPLC solvent 1 (isocratic): methanol/water 90:10 v/v, pH 6–7 (APCI positive and negative mode, ESI positive mode, ESI/APCI positive mode)
  - HPLC solvent 2 (isocratic): methanol/water 90:10 v/v, pH 8–9 (ESI negative mode, ESI/APCI negative mode)
  - Flow: 300 µL min^{-1}
  - Injection volume: 10 µL
  - Column: without column
- MS parameters:
  - MS: Agilent Technologies, triple quadrupole 6410
  - Ionization: multimode (ESI/APCI, positive/negative)
  - Drying gas temperature: 300 °C
  - Drying gas flow: 300 L h^{-1}
  - Nebulizer gas pressure: 30 psi
  - Vaporizer temperature: 200 °C
  - Capillary voltage: ($\pm$)2500 V (ESI/APCI, positive/negative)
  - Corona current: 7 µA (APCI positive/negative)
  - Charging voltage: 2000 V (ESI positive/negative)
  - Fragmentor voltage: 150 V (ESI/APCI, positive/negative)
  - Scan range: m/z 100–500 (9,10-anthraquinone); m/z 500–900 (malantide)

The experiments were carried out with either only ESI or APCI ionization mode or a combination of both in single runs each in positive or negative detection mode.

2.3   Methods

For choosing an appropriate ionization technique the most important parameters are knowledge of the behaviour and structure of molecules and also of the properties of the dissolving solutions. By considering both, the choice of the best-fitting ion source should no longer be a problem.

Figure 2.3 presents a flow injection experiment of the peptide malantide (left side) and the oxidized polycyclic aromatic hydrocarbon (oxy-PAH) 9,10-anthraquinone (right side) in various concentrations detected in positive ionization mode (Figure 2.3a–c) and in negative ionization mode (Figure 2.3d–f). The ionization was examined either by ESI (Figure 2.3a,d), by APCI (Figure 2.3b,e) or by ESI + APCI (Figure 2.3c,f). Suboptimal (default) settings were used for the ionization parameters, to ionize as universally as possible.

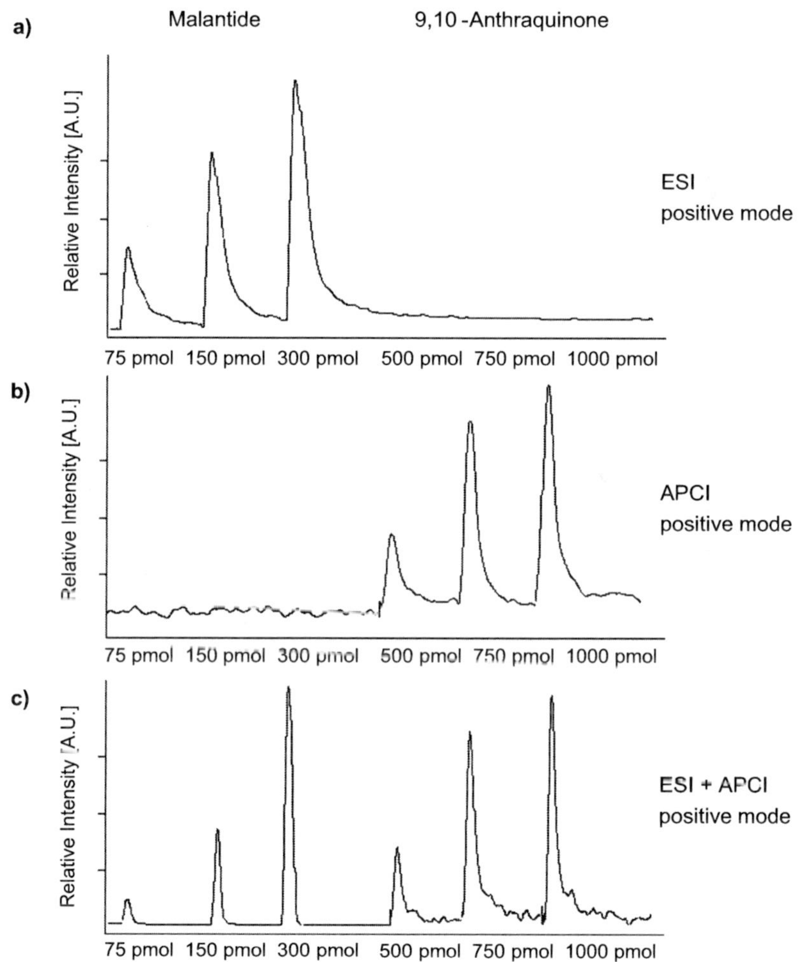

Figure 2.3 Flow injection experiment of the peptide malantide (left) and the oxy-PAH 9,10-anthraquinone (right) in various concentrations detected in positive ionization mode (a–c) and in negative ionization mode (d–f). The ionization was examined either by ESI (a, d), by APCI (b, e) or by MMI, *i.e.* ESI + APCI (c, f).

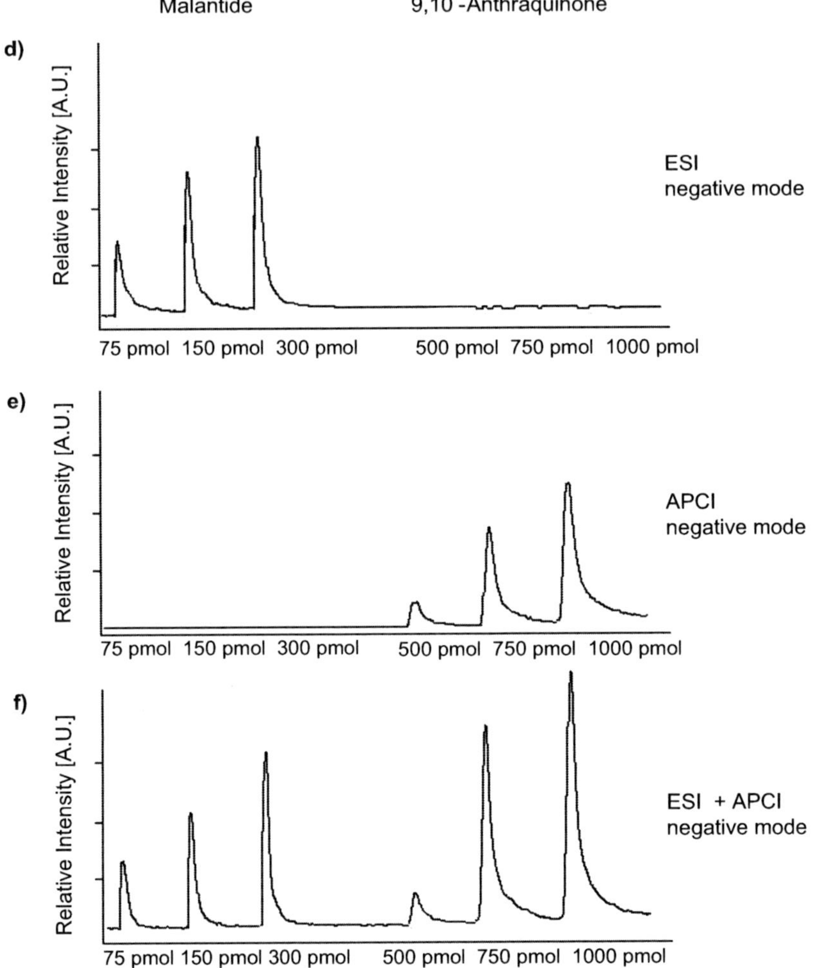

Figure 2.3   Continued

For maximum signal intensities to be achicved for each substance, the considerations set out in next section have to be considered.

2.3.1   Molecular Conditions

2.3.1.1   *Sprayed Analytes*

Generally, ESI is the first choice in spraying large biomolecules if the molecules:

- are charged at a specific pH-value due to
  - ➤ high proton affinity (like basicity on the protein surface; positive detection)

> ➤ or a strong proton desorption (like acidic groups have; negative detection),
- contains functional groups with weak binding strength to the molecule backbone (like protein glycosylations)
- typically have high molecular weight and can be multiply charged:
  - ➤ $m/z < 3000$ Da for normal mass spectrometers
  - ➤ higher m/z for high range mass spectrometer types (like ToF)
- can contain a higher amount of protons due to the higher content of basic groups on the protein surface (*e.g.* denatured protein)

Thus the 'biomolecule' malantide, a polypeptide, is effectively transferable from liquid phase to gas phase by an ESI source.

APCI (and APPI) is typically chosen for molecules that are uncharged in solution and with a high intramolecular binding strength. Thus oxy-PAH 9,10-anthraquinone, with an aromatic backbone, is effectively transferable from liquid phase to gas phase including ionization by an APCI source.

2.3.1.2 Sprayed Solvents and Additives

It is to essential consider the use of volatile salts and organic solvents in the spraying liquids. Especially for ESI (APCI is less sensitive to solvent) the spraying process is negatively influenced by so-called 'signal suppression'. Most problems are caused by the absence of an organic modifier and the presence of non-volatile compounds in the spraying solutions. Both result in ineffective droplet shrinking; the latter leads to a shift of the analytes' molecular weight by forming salt clusters and can also cause neutralization of analyte ions if salts are present in the millimolar range.

Also, large amounts of other additives like proteins, problematic acids (like trifluoroacetic acid) or reducing agents are not suitable for the same reasons mentioned above and additionally due to mass changing, complex formatting or ion neutralizing effects, all negatively influencing the detectability of gas-phase transferred ions. Figure 2.4 impressively reflects these effects on the detection of malantide sprayed with ESI in the presence of non-volatile salts and a large quantity of additives (Figure 2.4a) in comparison to that with MS-compatible conditions (Figure 2.4b). The solvent was applied also in a similar assay to that described in Chapter 10, in which it was essential to have MS-compatible conditions.

Further issues have to be kept in mind if one prepares the spraying solution:

- shifting pH values caused by analytes if no buffers are used, thus various ionization efficiency, thus missing stability for quantitative measurements
- signal suppression tests with matrix if real matrices are used
- concentrations of unavoidable non-volatile compounds, which should be kept as low as possible.[9,10]

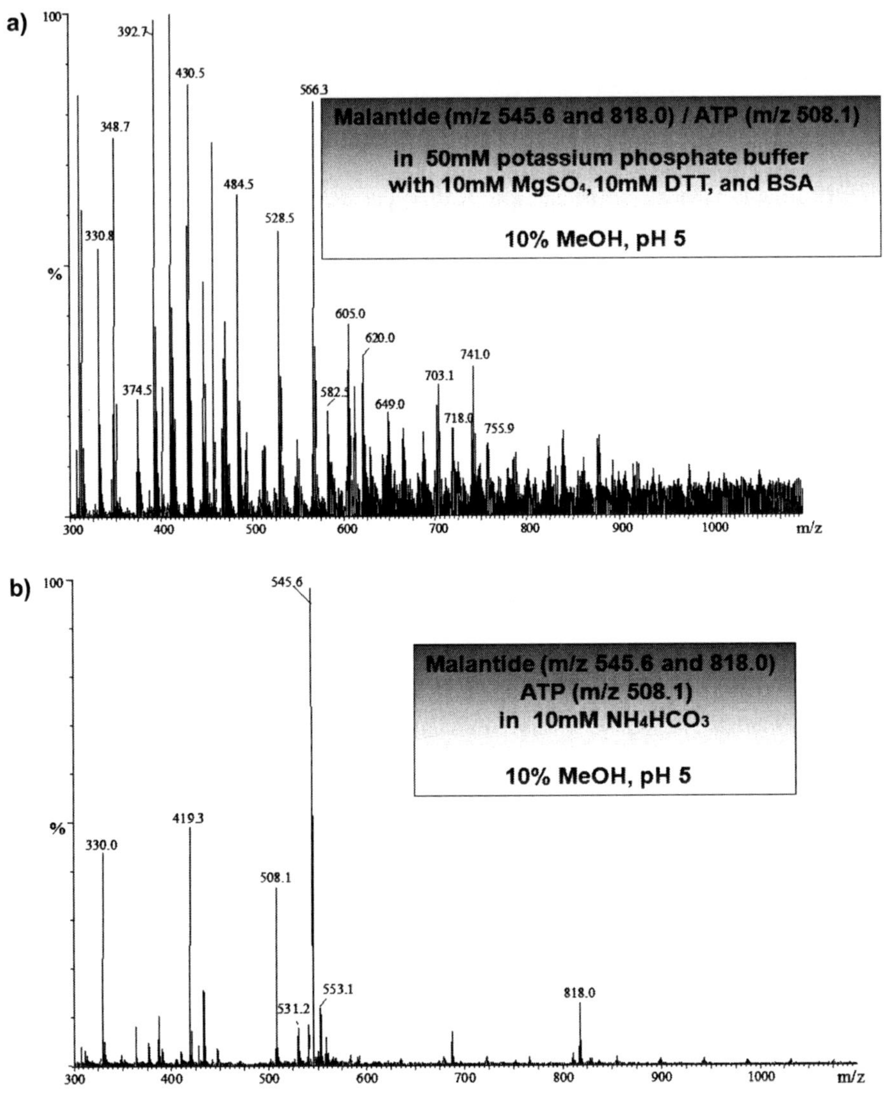

Figure 2.4   Mass spectra of the peptide malantide sprayed with ESI in presence of non-volatile salts and high amount of additives (a) and with MS-compatible conditions (b).

APCI is less influenced in all the mentioned respects. However, because of the relatively harsh conditions, APCI is only of limited utility for large biomolecules.

In addition, the chemical and physical behaviour of sprayed analytes, solvents and solvent additives must be considered and several instrumental optimization procedures have to be performed as well.

2.3.2 Spray Conditions (ESI)

Figure 2.2a shows the typical instrumental structure of an electrospray ion source and Figure 2.2b illustrates a newly developed ESI source for high solvent flow rates. An ESI source is principally constructed with a sprayer, an embedded needle connected *via* an inlet fitting to the HPLC outlet, an electric field, nebulizer gas and heated desolvation gas. The droplets are formed pneumatically and desolvated *via* heated nitrogen crossflow afterwards. The MS inlet is also often shielded by the gas flow. Ions transferred into gas phase can be transported *via* an electric field and the transfer line (*e.g.* the capillary in Figure 2.2a) into the mass analyser.

Several ion source parameters can or should be optimized. The most important ones are described in detail below.

> (i) Note that several suppliers of MS equipment use different terms for the source components and the associated source settings. I have tried to use easily identifiable names, but they will not always be the same as those used in your MS system

The **capillary voltage** induces an energetic field between the spraying needle and the MS inlet capillary. The capillary voltage typically can be set from 0.8 kV (if the spray capillary is very close to the MS inlet, as with nanospray needles) to 4.0 kV (if the spray capillary is very far from the mass spectrometric inlet, as with micro spray conditions above 100 µL min^{-1})

> (i) If the voltage is set too low, no ion transfer will occur into the MS inlet (Figure 2.2a, capillary); In contrast, if the chosen voltage is too high, the ion acceleration in this atmospheric pressure region will also be too high, resulting in a so called in-source fragmentation, *i.e.* a collision of ions with atmospheric gas-phase molecules. A voltage that is set too high often causes an electric arc between spraying capillary and MS inlet; be careful! Such a high voltage can severely damage your mass spectrometer!
>
> (i) In MS systems from Agilent systems, for example, the same is true for the so-called fragmentor region, an area within the mass spectrometer containing atmospheric gas molecules (region not shown in Figure 2.2a). Details can be found in Letzel *et al.*[11]

The **nebulizer gas pressure** is responsible for the correct droplet size while nebulizing the HPLC eluent. It can be set between 10 and 60 psi depending on the liquid flow (*e.g.* 1 µL min^{-1} and 200 µL min^{-1}, respectively).

> (i) If the pressure is too low, the primary droplets will be too large. Thus the drying process will be incomplete, and an ineffective ion transfer will be the result; if the pressure is too high, the ions will be transferred too early the in gas phase, thus an increased availability for in-source collision will

be the result and an increased loss of molecules by in-source fragmentation will occur

- ⓘ If the sprayer position is far from the MS inlet, the pressure has to be higher than in the case of a sprayer position closer to the MS inlet, independently of the resulting droplet size

The **desolvation gas temperature** and **flow** are critical, because of the reverse gas flow (see Figure 2.2a, 'heated N2') which supports the droplet shrinking.

- ⓘ If the temperature and/or flow is too low the droplets cannot be dried, thus less ion transfer into the gas phase and hence into the MS can be obtained. If both are set too high the molecules will be transferred into the gas phase too early, thus getting into direct contact to hot air which can cause molecule fragmentation

To optimally transfer ions, experience in choosing the right settings is needed. However if one keeps these tips in mind, results of good quality will come soon. Malantide can be ionized very sensitively in protonated form (positive detection mode), less so in deprotonated form (negative detection mode), whereas 9,10-anthraquinone cannot be ionized at all, although it is possible with an APCI source.

2.3.3 Spray Conditions (APCI)

Figure 2.2c shows a typical structure of an APCI source with important system parts, like the high liquid flow sprayer with embedded needle connected *via* a fitting inlet to the HPLC outlet and pressurized by vaporized gas. The vapour droplets are desolvated by cross-flow heated nitrogen, later also shielding the MS inlet. Ions are produced in the corona needle region. Subsequently the ions are transferred into the mass analyser by an electric field and transfer line (*e.g.* the capillary in Figure 2.2c). The APCI corona needle provides electrons for charging molecules, whereas the APPI source provides photons (not used in this study, see Figure 2.2d). Both mechanisms are described in detail in the literature.[2,3]

Additionally to the source parameter capillary voltage, nebulizer gas pressure, vaporizer gas temperature, desolvation gas temperature and flow already described in the previous section on ESI, two further ion source parameters that have to be optimized are the corona current and the vaporizer gas temperature.

The **corona current** is responsible for the amount of electrons available for the charging effect.

- ⓘ The default settings can usually be used for the parameter corona current; however, higher and lower values should always be tested in order to optimize ionization efficiency and specificity

The corona current ionizes the analytes by collision with ionized solvent intermediates. The latter are in excess in solution and are preferably ionized in the first step. Details of the APCI mechanism can be found in a recently published review.[2] The ionization mechanism of the similar ion source APPI can be found in another review.[3]

The **vaporizer gas temperature** supports a sufficient spraying process, but is normally not too critical because of the heat capacity of the excess liquid molecules.

An increase of ionization efficiency of non-polar molecules can be increased by adding solvent additives, called dopants, that should be present prior to the spraying process. These additives (*e.g.* acetone or dichloromethane) are not suitable for HPLC separation, thus they have to be added in a postcolumn mixing flow.[12]

Compared with ESI, APCI is more matrix independent, thus non-volatile salt systems can be used.

2.3.4 Spray Conditions (MMI; *i.e.* ESI + APCI)

A recently developed ionization source is multimode ionization (MMI) combining an ESI interface with an APCI interface (see Figure 2.2e). This source incorporates all of the advantages and disadvantages of both interfaces and has therefore to be optimized as a compromise between the two. However, the source used in this study has an IR emitter as drying origin and not a radiative control by heated gas. Thus, the conditions are harsher than in classical APCI (in which the heat of the drying gas, and thus the influencing energy, can easily be regulated). This source features an IR emitter of about 10 cm in length that can be regulated in strength but not in space. Thus the molecules are partially destroyed or converted, which leads to a lower significant ion intensity in the mass spectrometer and also causes greater contamination in the ion source. A hood or classical drying gas would overcome this problem and would also be useful to improve the transfer, although the observation of proteins is more effective without destruction or change of drying gas.

Again, the source is less matrix dependent than ESI.

2.4 Notes

- Be aware that each supplier uses a different nomenclature for their own ion source settings (*e.g.* Agilent in this chapter, Waters in Chapter 11, and Thermo Fisher Scientific in Chapter 9)
- Think about the molecules to be analysed; consider molecular structure (*e.g.* heteroatoms or aromatic backbone), polar groups (*e.g. via* proton affinity), and intramolecular binding strength to choose the appropriate ionization technique useful for your conditions and analytes
- Think about the analyte desolving and spraying solvents; consider viscosity, pH value, analyte solvation energy, volatility, complex-forming

properties, heat capacity, surface tension of formed droplets, initial droplet size (nano or micro),[13,14] and then you will know if you can get an effective ion transfer into your mass spectrometer

- Think about the solution ingredients like salts, acids, proteins and other additives; consider the amount and nature of molecules that can influence the spraying and ionization process negatively by forming complexes, slowing down the droplet shrinking, neutralizing gas-phase ions, often consuming protons in the gas phase. Typically for ESI spraying process the total amount of non-volatile additives should not be in the millimolar region, whereas for APCI and APPI it can be significantly higher
- Think about the technical parameters; how high is the liquid flow rate, what does the solution look like? Thus the capillary voltage, the position of the spraying 'capillary', the diameter of the spraying 'capillary', spraying conditions such as pressure and temperature of spraying gases, electron energy (APCI) or photon energy (APPI) or dopants (APCI and APPI) have to be optimized for the most effective ion transfer
- Think about the type and supplier of the mass spectrometer; not every system is optimized in the ionization source region (especially in the use of APCI and APPI). However, if you already have a machine in your lab, it is too late, because the tests and choice have to be made before the equipment is bought
- Last but not least, think about a good chromatographic separation (prior spraying) and correct MS conditions in the analyzer (after spraying). This leads to a successful analysis procedure with regard to LC-API-MS with the ion source as the key element

I anticipate that this can help you in performing and generating optimum results in identifying, quantifying and monitoring your molecules of interest. Viel Erfolg! (German for: good luck!)

References

1. J. B. Fenn, M. Mann, C. K. Meng, S. F. Wong and C. M. Whitehouse, *Science*, 1989, **246**(4926), 64.
2. E. Rosenberg, *J. Chromatogr., A*, 2003, **1000**, 841.
3. S. J. Bos, S. M. Van Leeuwen and U. Karst, *Anal. Bioanal. Chem.*, 2006, **384**, 85.
4. H. Chen, A. Wortmann, W. Zhang and R. Zenobi, *Angew. Chem, Int. Ed.*, 2007, **46**, 580.
5. S. Sakamoto, M. Fujita, K. Kim and K. Yamaguchi, *Tetrahedron*, 2000, **56**, 955.
6. R. E. Ardrey, *Liquid Chromatography-Mass Spectrometry: An Introduction*, John Wiley & Sons Chichester, 2003.
7. C. Dass, *Principles and Practice of Biological Mass Spectrometry*, John Wiley & Sons, New York, 2001.

8. C. Berkemeyer and T. Letzel, *LC-GC Europe*, 2008, **21**, 548.
9. M. C. García, A. C. Hogenboom, H. Zappey and H. Irth, *J. Chromatogr., A*, 2002, **957**, 187.
10. A. R. de Boer, T. Letzel, H. Lingeman and H. Irth, *Anal. Bioanal. Chem.*, 2005, **381**, 647.
11. T. Letzel, U. Pöschl, E. Rosenberg, M. Grasserbauer and R. Niessner, *Rapid Commun. Mass Spectrom.*, 1999, **13**, 2456.
12. S. Grosse and T. Letzel, *J. Chromatogr. A*, 2007, **1139**, 75.
13. P. Kebarle and L. Tang, *Anal. Chem.*, 1993, **65**, A972.
14. P. Kebarle and U. H. Verkcerk, *Mass Spectrom. Rev.*, 2009, **28**, 898.

Expression and Purification of Bioactive Proteins/Peptides with Conventional Liquid Chromatography

TAKAYUKI OHNUMA AND TAMO FUKAMIZO

Kinki University, Department of Advanced Bioscience,
3327–204 Nakamachi, Nara, 631–8505, Japan

3.1 Introduction

Escherichia coli (*E. coli*) is one of the most widely used hosts for the production of recombinant proteins/peptides. It is easily manipulated and can grow quickly in inexpensive media. Since an *E. coli* cell can accumulate recombinant proteins/peptides up to 80% of its dry weight, this host cell is suited for massive production of proteins/peptides.

The strategy for expressing recombinant proteins/peptides in *E. coli* starts with the construction of the expression vector. This step involves insertion of the gene encoding proteins/peptides of interest into an expression vector, usually a plasmid. The expression vector needs to have the following features:

- an origin of replication to control the plasmid copy number
- a gene encoding selectable marker to maintain the vector in the cell

RSC Chromatography Monographs No. 15
Protein and Peptide Analysis by LC-MS: Experimental Strategies
Edited by Thomas Letzel

Published by the Royal Society of Chemistry, www.rsc.org

- a controllable transcriptional promoter (*e.g.* T7,[1] *lac*[2] or *araBAD*[3]) to initiate transcription
- a transcriptional terminator
- a ribosome-binding sequence for initiation of translation

Next, the expression vector containing the gene to be expressed is introduced into an appropriate *E. coli* strain by transformation. Expression of the proteins/ peptides in the cells can be induced by the addition of the appropriate inducer (*e.g.* IPTG, lactose or arabinose) into culture when the cells are growing at mid log phase. After induction, the cultures are incubated from 3 hours to overnight depending on the induction temperature (3 h at 37 °C and overnight below 20 °C) for the expression.

Ion exchange column chromatography has frequently been used for the purification of proteins/peptides. In principal, it is based on charge–charge interactions between the surface of proteins/peptides and the ionizable functional groups immobilized on the stationary phase (resin).[4] There are two types of ion exchange column chromatography: cation exchange column chromatography and anion exchange column chromatography. In anion exchange column chromatography, negatively charged ions (proteins/peptides) bind to a positively charged resin. Conversely, in cation exchange, ions in proteins/ peptides need to be positively charged, and the resin is negatively charged. The commonly used functional groups are shown in Table 3.1. The diethylaminoethyl (DEAE) group is a weak base that has a net positive charge when ionized and therefore binds and exchanges negatively charged ions. The carboxymethyl (CM) group is a weak acid that has a negative charge when ionized and binds and exchanges positively charged ions. Other ion exchange resins commercially available carry sulfopropyl (SP) and methyl sulfate (S) for cation exchange and a quaternary amino ethyl (QAE) and quaternary ammonium (Q) for anion exchange.

Gel filtration column chromatography is a method to separate proteins/ peptides based on differences in their hydrodynamic radius, which correlates well to molecular weight.[5] It employs a column packed with porous beads, ideally of neutral surface chemistry. A large number of gel filtration resins with a wide range of pore sizes are commercially available (see also Table 3.2).

Table 3.1 Functional groups used for ion exchange column chromatography

Functional group	*Structure*	*Nature*
Anion exchangers		
Diethylaminoethyl (DEAE)	$-N^+(C_2H_5)_2H$	Weak anion
Quaternary amino ethyl (QAE)	$-CH_2CH_2N^+(CH_3)_3$	Strong anion
Quaternary ammonium (Q)	$-CH_2N^+(CH_3)_3$	Strong anion
Cation exchangers		
Carboxymethyl (CM)	$-O-CH_2COO^-$	Weak cation
Sulfopropyl (SP)	$-CH_2CH_2CH_2SO^-$	Strong cation
Methyl sulfate (S)	$-CHOHCH_2SO^-$	Strong cation

Although large differences in separation are realized, the variation in the pore sizes of these various beads is actually in a fairly narrow range. When the proteins/peptides are applied to the column the smaller proteins/peptides relative to the pore size enter the pores in the beads as the working buffer flows. Intermediate-size proteins/peptides partially or only occasionally enter the beads, depending on the size and shape of the molecules and individual pores. Migration of these molecules in the column is retarded by their repetitive penetration into the beads, which makes their path to the end of the column longer. In contrast, larger proteins/peptides are excluded from the pores altogether and remain in the mobile phase. Eventually, the proteins/peptides applied to the column are eluted in an inverse order of their approximate molecular weight.

Hydrophobic interactions are the most important non-covalent force that is responsible for a wide variety of biological phenomena, such as stabilization of the three-dimensional structure of proteins, substrate binding to enzymes, folding of proteins, and antibody–antigen reactions. Proteins/peptides can bind to a hydrophobic resin by this interaction at the non-polar regions of their surface. Of the 20 amino acids that constitute proteins/peptides, 8 are classified as hydrophobic and are primarily responsible for these interactions, *via* their non-polar side chains. Proteins/peptides possess different numbers and kinds of hydrophobic amino acid residues on their surface, giving rise to different degrees of interaction force.

A variety of resins for hydrophobic interaction column chromatography are also commercially available (Table 3.3). The extent of hydrophobicity depends

Table 3.2   Gel filtration media

Medium	*Chemistry*
Sephadex	Dextran cross-linked with epichlorohydrin
Sephacryl	Allyl dextran cross-linked with N,N'-methylenebisacrylamide
Sepharose	Agarose
Superdex	Composite of cross-linked agarose and dextran
BioGel A	Agarose
BioGel P	Acrylamide cross-linked with N,N'-methylenebisacrylamide
Ultrogel A	Agarose
Ultrogel AcA	Composite of agarose and polyacrylamide

Table 3.3   Functional groups used for hydrophobic interaction column chromatography

Functional group	*Structure*	*Hydrophobicity*
Ether-	$HO(CH_2CH_2O)_n-$	
Phenyl-	C_6H_5-	
Butyl-	C_4H_9-	$\downarrow$
Hexyl-	$C_6H_{13}-$	
Octyl-	$C_8H_{17}-$	
Decyl-	$C_{10}H_{21}-$	Strong

on the number of CH_2 groups attached to the beads and their flexibility of the hydrocarbon chain.[6] In hydrophobic interaction chromatography, proteins/ peptides are applied to the resin in a high-salt buffer (ammonium sulfate is the most popular) and elution is achieved by a descending salt gradient. Note that despite some similarities to RP-LC (reversed-phase liquid chromatography), hydrophobic interaction chromatography is quite distinct. Since RP-LC is generally performed with the mobile phase containing organic solvent, it is not typically used for protein separations, because the organic solvent denatures many proteins and inactivates their biological function.

Affinity chromatography separates proteins/peptides based on a reversible interaction between proteins/peptides and their specific ligands immobilized on the resins. This type of chromatography relies on biological function. Since the interaction is not due to general properties such as pI, hydrophobicity, and molecular weight, it often enables a highly selective separation.[7] Two classes of ligands are used for affinity chromatography, monospecific and group-specific ligands. Monospecific ligands interact with only a single protein or a very small number. These interactions can be seen between receptors, antibodies and enzymes and their respective naturally occurring ligands or derivatives thereof (*i.e.* substrates or inhibitors in the case of enzymes). Group-specific ligands include enzyme cofactors, such as NAD^+ and $NADP^+$, for isolation of the NAD^+ and $NADP^+$-dependent enzymes: lectins for glycoproteins, proteins A and G for IgG, and calmodulin for calcium-dependent enzymes. For elution of bound proteins or peptides from affinity resins, the free ligand in working buffer, which competes with the immobilized ligand for the protein's binding site, is often used. Alternatively, non-specific methods such as high salt, changing pH and/or temperature, and even cold water, which minimizes hydrophobic interactions, can be used. On the other hand, it should be noted that affinity chromatography increasingly uses a number of protein tags genetically fused onto recombinant proteins. This approach has become a powerful method for protein and peptide purification. Affinity tags can be broadly classified into two categories, small peptide tags and large peptide/protein tags (Table 3.4). Small tags sometimes do not need to be cleaved out from the fusion proteins because they are small enough not to interfere with protein conformation, disrupt the function of the bait and/or be immunogenic. They include six-polyhistidine

Table 3.4 Affinity tags and resins

Tag	*Sequence*	*Resin*
Small tags		
Poly-His	HHHHHH	Ni-NTA
Strep-tag II	WSHPQFEK	Strep-Tactin
FLAG	DYKDDDDK	Anti-FALG antibody
S-	KETAAAKFERQHMDS	S-fragment of RNase A
Large tags		
Chitin-binding domain		Chitin
Maltose-binding protein		Cross-linked amylose
Glutathione S-transferase		Glutathione

residues (His), Strep-tag II and FLAG. Larger tags may not only facilitate the purification procedure of the fusion proteins, but also sometimes enhance the expression level and solubility. They can be removed by chemical agents or by enzymatic means (protease digestion), or intein-mediated splicing. They include chitin binding protein (CBP), maltose binding protein (MBP), thioredoxine (Trx) and glutathione-*S*-transferase (GST).[8]

3.2   Experimental

3.2.1   Materials

- pET-Blue-1 and Tuner(DE3) pLacI were purchased from Novagen (Madison, USA)
- SP Sepharose FF, HiPrep 26/60 Sephacryl S-100 HR, and Hitrap Phenyl HP were purchased from GE Healthcare Life Sciences
- Sepasol-RNAI was from Nacalai Tesuque Inc. (Kyoto, Japan)
- DNase I and ReverTra Ace were obtained from Takara Shuzo (Kyoto, Japan) and Toyobo (Osaka, Japan), respectively

3.2.2   FPLC System

- FPLC system, Amersham Biosciences, Uppsala, Sweden

3.2.3   Cloning cDNA and Construction of the Expression Vector

- Extract total RNA from tobacco leaves (1.0 g fresh weight) grown for 5 weeks using Sepasol-RNAI
- To eliminate any DNA, treat the RNA with DNase I and convert it into cDNA using the ReverTra Ace with the oligo(dT)20 primer
- Amplify the cDNA encoding mature chitinase protein (NtChiV) from the first strand cDNA using the forward primer, 5′-ATGCAAAATGTTAAG-GGAGGATACTGGT-3′, and reverse primer, 5′-TTACTTCATCTC-TTGAAATGACACTCCCCA-3′, designed from the genomic sequence of NtChiV.[9]
- Ligate the PCR products into pET-Blue-1 vector by TA-cloning
- Confirm the cDNA sequence of NtChiV coding gene region of the resulting plasmid (pETB-NtChiV)

3.2.4   Recombinant Protein Expression in *E. coli*

- Introduce pETB-NtChiV into *E. coli* Tuner(DE3) pLacI
- Grow *E. coli* Tuner(DE3) pLacI harbouring the recombinant expression plasmid to $A_{600} = 0.8$–1.0
- Add inducer IPTG to the culture to a final concentration of 1 mM
- Continue to grow the cells for 20 h at 18 °C
- Harvest the cells by centrifugation at 5000 *g* for 15 min at 4 °C

3.2.5 Separation by Ion Exchange Chromatography

- Suspend the cells in a 10 mM sodium phosphate buffer, pH 7.5, and disrupt with a sonicator
- Remove cell debris by centrifugation at 20 000 g for 20 min at 4 °C
- Dialyse the resulting supernatant against the same buffer
- Eliminate the resulting insoluble proteins by centrifugation at 20 000 g for 15 min
- Apply the supernatant on to a 1 × 5 cm SP Sepharose FF column equilibrated with the dialysis buffer
- Wash the column with 30 mL of the same buffer
- Elute the adsorbed proteins with a linear gradient of NaCl from 0 to 0.3 M in the same buffer

3.2.6 Separation by Gel Filtration Chromatography

- Pool the fractions containing chitinase activity eluted from the SP Sepharose chromatography column
- Concentrate the protein solution to 5 mL by Amicon filter device (3000 MWCO)
- Apply to a HiPrep 26/60 Sephacryl S-100 HR previously equilibrated with 10 mM sodium phosphate buffer containing 0.1 M NaCl, pH 7.5 and developed with the same buffer using an FPLC system

3.2.7 Separation by Hydrophobic Interaction Chromatography

- Pool the fractions containing chitinase activity eluted from the gel filtration chromatography column
- Mix with an equal volume of 10 mM sodium phosphate buffer containing 2 M ammonium sulfate, pH 7.5, and apply to a Hitrap Phenyl IIP (1 × 1.6 cm) previously equilibrated with 10 mM sodium phosphate buffer containing 1 M ammonium sulfate, pH 7.5
- Wash the column with the same buffer with 1 M ammonium sulfate
- Elute the adsorbed proteins with a linear gradient of ammonium sulfate from 1.0 to 0 M using an FPLC system

3.3 Results and Discussion

3.3.1 Ion Exchange Chromatography

In ion exchange column chromatography, almost all proteins that were of host *E. coli* origin, except for NtChiV, passed through the column without binding to the resin (Figure 3.1). Under the experimental conditions, NtChiV bound to the resin and eluted at 0.05 M NaCl concentration. As judged by SDS-PAGE, a major protein band with a molecular weight of 39 000, corresponding to the

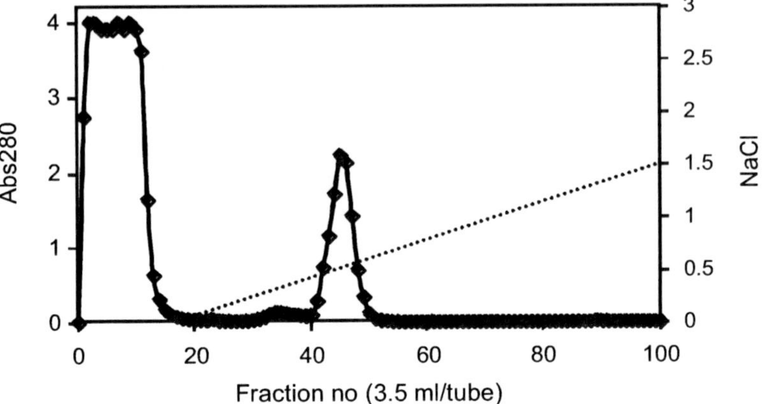

Figure 3.1   SP Sepharose FF column chromatogram of crude extract from *E. coli* cells.

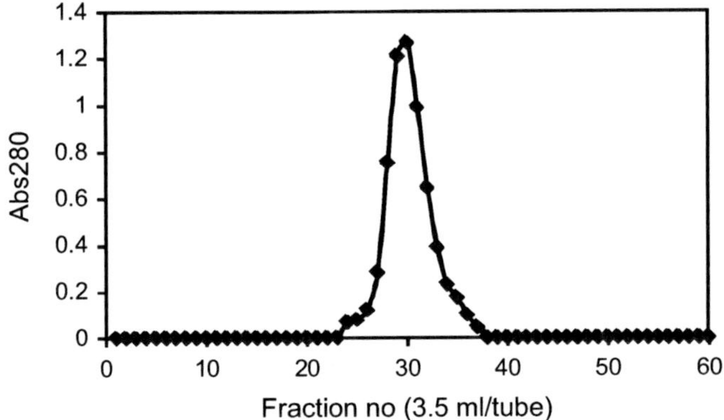

Figure 3.2   HiPrep Sephacryl S-100 HR column chromatogram of chitinase fractions obtained by ion exchange on a SP Sepharose FF column.

theoretical weight of NtChiV (39 033.68), and a few minor protein bands with lower molecular weights were observed (Figure 3.4, lane 2). This step was the most effective for eliminating proteins of bacterial origin.

3.3.2   Gel Filtration Chromatography

The elution profile of NtChiV on HiPrep Sephacryl S-100 HR is shown in Figure 3.2. One symmetrical protein peak with chitinase activity was eluted at 30 mL of elution volume, and the peak fraction still showed a few minor protein bands along with the major NtChiV protein band on SDS-PAGE by CBB staining (Figure 3.4, lane 3).

3.3.3 Hydrophobic Interaction Chromatography

The elution profile of NtChiV on Hitrap Phenyl HP is shown in Figure 3.3. One symmetrical protein peak with chitinase activity eluted at 0.5 M ammonium sulfate concentration was obtained and this peak gave a single band on SDS-PAGE (Figure 3.4, lane 4). Proteins that copurified with NtChiV by ion exchange and gel filtration chromatography were separated from NtChiV by this procedure.

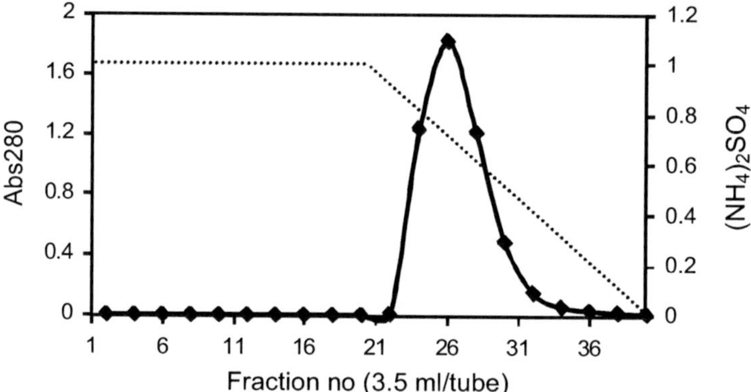

Figure 3.3 Hitrap Phenyl HP column chromatogram of chitinase fractions obtained by gel filtration on a Sephacryl S-100 HR column.

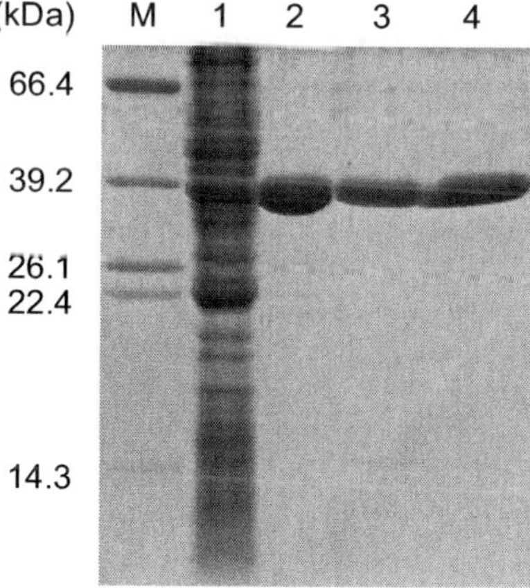

Figure 3.4 SDS-PAGE showing purification of recombinant NtChiV. Lane 1, crude extract; lane 2, pooled proteins after SP Sepharose chromatography; lane 3, pooled proteins after Sephacryl S-100 HR chromatography; lane 4, pooled proteins after Hitrap Phenyl HP chromatography.

Table 3.5   Purification of NtChiV from *E. coli*

Step	*Protein (mg)*	*Activity ($\mu mol\ min^{-1}$)*	*Specific activity ($\mu mol\ min^{-1}\ mg^{-1}$)*	*Yield (%)*
Crude extract	2982.0	159.4	0.05	100
SP Sepharose	53.24	62.3	1.17	39.1
Sephacry S-100 HR	28.76	31.3	1.20	21.6
Hitrap Phenyl HP	14.73	34.4	2.12	19.6

3.3.4   Overall Evaluation

The enzyme was successfully purified to homogeneity (a single protein band on SDS-PAGE), corresponding to a 42.4-fold increase in specific activity compared to the crude enzyme, and a 20% yield. The purification procedures are summarized in Table 3.5. The yield of recombinant NtChiV was about 15 mg from 0.6 L of induced culture.

3.4   Notes

3.4.1   Other Hosts for Expression

One of the most useful systems for expression of recombinant proteins/peptides in *E. coli* is the pET vector series (Novagen), which is based on the T7 phage RNA polymerase promoter and uses the pBR322 origin of DNA replication. The expression of the recombinant proteins/peptides using these plasmids is tightly regulated, and, when induced, produces high levels of transcripts and recombinant proteins/peptides.[1] If the *E. coli* expression system fails to produce soluble, functional protein, alternative strategies should be considered. The *Brevibacillus* host (Takara) is one such alternative strategy.[10] This strain produces large amounts of proteins and secretes them into culture media. Eukaryotic expression systems, such as baculovirus expression system in insect cells,[11] a yeast expression system, such as *Pichia pastoris* (Invitrogen), the most widely used strain,[12] *Kluyveromyces lactis* (New England Biolabs), and human cells,[13] are other alternatives. For proteins that require glycosylation, the yeast, baculovirus, or mammalian cell system should be used. Cell-free systems with prokaryotic and eukaryotic extracts also are promising, especially for toxic proteins to the desired host strains.[14] They have been used to generate a number of proteins for structural studies.

3.4.2   Ion Exchange Chromatography

The ampholytic nature of proteins has to be considered upon choosing the ion exchanger and working buffer. Anion exchange chromatography should be carried out above the pI of the target proteins/peptides and a cation exchange below the pI to make them have the opposite charge to the resins. However, it has often been observed that proteins/peptides can bind to the resin at a variety of pH values since the local charge density of the exposed surface influences their chromatographic behaviour on ion exchange chromatography. If the pIs of proteins/peptides of interest are known, the pH of the working buffer is usually adjusted to at least 1 pH unit above the pI for an anion exchange or at least 1 pH unit below the pI for a cation exchange column chromatography, for better binding. In cases where the pIs of the proteins/peptides are unknown, ion exchange column chromatography with both anion and cation exchangers may need to be conducted until the conditions giving good resolution are established. Several molecular structures of ion exchange column materials can be found in Table 3.1. The DEAE type resins are usually used in the pH rage 4–9. Since the pK_a value of the DEAE group is about 9.5, it shows good ion binding capacity below a pH of about 8.5. If anion exchange column chromatography needs to be performed at higher pH, the QAE type resins can be used since they are completely ionized over a wide pH range (pH 2–12). Similarly, the CM type resins are used in the pH rage 4.5–9.5 (pK_a of about 4) and the SP type in the pH 4–12 range. To select the working buffer correctly, one should consider that for anion exchange column chromatography, cationic buffers, such as Tris, alkylamines, ammonium, imidazole, ethyldiamine, pyridine, and aminoethyl alcohol, are best, since the opposite charge interferes with the ion exchange process and causes local disturbances in pH. Similarly, anionic buffers, such as phosphate, acetate, citrate, glycine, and barbiturate, should be used for cation exchange column chromatography. Elution of proteins/peptides bound to the resin can be achieved by varying either the pH or the ionic strength of the elution buffer. At the starting point of the chromatography, the ionic strength of the working buffer should be set at a lower level in order to maintain the charge–charge interaction between proteins/peptides and the resin and enhance binding. By increasing the NaCl concentration in the buffer, the salt ion in the buffer starts to compete with the proteins/peptides for binding to the resin. Eventually, bound proteins/peptides no longer interact substantially with the resin and are released at a given concentration.

3.4.3 Gel Filtration Chromatography

Three important parameters must be considered when choosing a gel filtration resin—V_o, V_e and V_t. These parameters must be determined experimentally. V_o is the total volume of the fluid that occupies the space between the beads of stationary phase, also called the void volume. V_t is the total volume, *i.e.* the sum of V_o and the volume of the fluid existing within the pores of the beads. V_e is the elution volume of the molecule of interest, which should be intermediate between V_o and V_t. The molecular weight is not the only factor that affects the chromatographic behaviour of the molecules, but ultimately a resin that has a smaller value of V_o and a larger value of V_t than V_e needs to be used to obtain a sufficient separation. Commonly used gel matrices are cross-linked products of dextran (Sephadex), agarose (Sepharose and Bio-Gel A) and polyacrylamide (Bio-Gel P). By changing the degree of cross-linking, these matrices acquire different degrees of porosity, and thus are able to fractionate proteins/peptides of a wide molecular weight range. The Sephadex resin is chemically stable in water, salt solution, organic solvents, and buffers of a wide pH range. However, Sephadex G-100 and G-200 resins, which have relatively larger pore size, are physically fragile and tend to be easily compressed even under low pressure during the chromatographic process. For this reason, they are not suitable for large-scale purification. To avoid such inconvenience, the Sephacryl and Ultrogel resins are now widely used. These two resins are cross-linked products of allyl dextran with N,N'-methylene bisacrylamide and agarose with acrylamide, respectively, and are physically more stable than the conventional Sephadex resins. Sepharose is made of the polysaccharide agarose. This resin has relatively large pore size, which makes it suitable for fractionation of high molecular weight proteins. The Sepharose resin, however, is not physically stable and thus not useful for production on the industrial scale. Superdex is a resin that combines the excellent gel filtration properties of dextran with the physical and chemical stability of highly cross-linked agarose. This resin withstands under high pressure, allowing fast flow in the chromatographic elution, hence reducing the total time for gel filtration chromatography.

3.4.4 Hydrophobic Interaction Chromatography

Resins possessing phenyl groups are most frequently used as an adsorbent for hydrophobic interaction column chromatography. The phenyl group is intermediate between n-butyl and n-pentyl in hydrophobicity, and bind to aromatic amino acids through π–π interactions. Resins possessing octyl groups should be used only for weakly hydrophobic proteins/peptides. The hydrophobicity of the octyl group is so strong that it is often difficult to elute bound protein/peptide from the resin without protein denaturation. In hydrophobic interaction column chromatography, samples are applied to the resin in a high-salt buffer (ammonium sulfate is the most popular) and elution is achieved by a descending salt gradient. HILIC (hydrophilic interaction liquid chromatography) is a variant of normal phase chromatography that is performed with a very polar stationary phase and a hydrophobic (mostly organic) mobile phase.[15] HILIC can be used for the separation of proteins, peptides, amino acids, oligonucleotides and carbohydrates and is especially good at separating hydrophilic peptides. It should be noted that HILIC mobile phases are compatible with ESI-MS, making this separation mode a viable alternative.[16] A detailed description of the HILIC-MS coupling and other MS-compatible separation techniques can be found in Chapter 4.

References

1. F. W. Studier, A. H. Rosenberg, J. J. Dunn and J. W. Dubendorff, *Methods Enzymol.*, 1990, **185**, 60.
2. B. Gronenborn, *Mol. Gen. Genet.*, 1976, **148**, 243.
3. L. M. Guzman, D. Belin, M. J. Carson and J. Beckwith, *J. Bacteriol.*, 1995, **177**, 4121.

4. E. F. Rossomando, *Methods Enzymol.*, 1990, **182**, 309.
5. E. Stellwagen, *Methods Enzymol.*, 1990, **182**, 317.
6. J. T. McCue, *Methods Enzymol.*, 2009, **463**, 405.
7. S. Ostrove, *Methods Enzymol.*, 1990, **182**, 357.
8. K. Terpe, *Appl. Microbiol. Biotechnol.*, 2003, **60**, 523.
9. L. S. Melchers, M. Apotheker-de Groot, J. A. van der Knaap, A. S. Ponstein, M. B. Sela-Buurlage, J. F. Bol, B. J. Cornelissen, P. J. van den Elzen and H. J. Linthorst, *Plant J.*, 1994, **5**, 469.
10. M. Mizukami, H. Hanagata and A. Miyauchi, *Curr. Pharm. Biotechnol.*, 2010, **11**, 251.
11. T. A. Kost, J. P. Condreay and D. L. Jarvis, *Nat. Biotechnol.*, 2005, **23**, 567.
12. J. M. Cregg, J. L. Cereghino, J. Shi and D. R. Higgins, *Mol. Biotechnol.*, 2000, **16**, 23.
13. M. P. Rosser, W. Xia, S. Hartsell, M. McCaman, Y. Zhu, S. Wang, S. Harvey, P. Bringmann and R. R. Cobb, *Protein Expr. Purif.*, 2005, **40**, 237.
14. L. Jermutus, L. A. Ryabova and A. Plückthun, *Curr. Opin. Biotechnol.*, 1998, **9**, 534.
15. P. Hemström and I. Knut, *J. Sep. Sci.*, 2006, **29**, 1784.
16. H. P. Nguyen and K. A. Schug, *J. Sep. Sci.*, 2008, **31**, 1465.

Liquid Chromatography-Mass Spectrometry of Intact Proteins

NICOLAS L. YOUNG[1] AND BENJAMIN A. GARCIA[1,2]

[1] Princeton University, Department of Molecular Biology, Princeton, NJ 08544, USA; [2] Princeton University, Department of Chemistry, Princeton, NJ 08544, USA

4.1   Introduction

The liquid chromatography-mass spectrometry (LC-MS) analysis of intact proteins is a significantly less common approach in proteomic analysis than the LC-MS analysis of protein proteolysis products, as in the ubiquitous LC-MS 'bottom up' mass spectrometric analysis (see also Chapter 1). This is because the LC-MS of intact proteins presents multiple challenges on both the LC and the MS sides of this technique. These challenges derive from the size and structural flexibility of proteins. Most experts consider the LC side of the challenge greater. There is no simple universal approach to protein chromatography and little progress has been made or is expected in this regard.[1] On the other hand, the capacity to analyse large intact proteins by MS has progressed rapidly and has become increasingly available in the past few years. In addition, substantial protein sample preparation challenges, such as solubility, persist throughout the analysis. Most who work in the field of proteomics work with peptides and use nearly uniform octadecylsilane (C18)-based separations and collision induced dissociation (CID) of typically doubly charged tryptic peptides for protein identification based on sequence tags of component peptides. This does not prepare one well for intact protein

RSC Chromatography Monographs No. 15
Protein and Peptide Analysis by LC-MS: Experimental Strategies
Edited by Thomas Letzel

Published by the Royal Society of Chemistry, www.rsc.org

LC-MS and can give a naive sense of confidence in approaching yet another project. Intact protein LC-MS is most often targeted at one protein of interest or a limited class of proteins, and is never as large scale as bottom up LC-MS analysis. Because of the extensive challenges, many intact protein LC-MS methods do not provide as complete characterization of the proteins detected as one might be imagine to be possible. However, intact protein LC-MS excels at certain types of information not otherwise available. For example, single nucleotide polymorphisms are more confidently identified by such approaches. Similarly, the identification of novel splice variants and their relative ratios are near impossible by other means (save top down mass spectrometry without the online LC component). In this chapter we introduce some of the fundamental challenges and solutions of on-line LC-MS of intact proteins (off-line LC followed by MS will not be included). Figure 4.1 gives an overview of a generic protein LC-MS analysis from far upstream to final results. We will focus primarily on the later steps in this process that are unique to protein LC-MS.

4.2 Liquid Chromatography

Proteins, as a class, are not amenable to any single, uniform physical separation methodology. Proteins are diverse in physical character from protein to protein (from very acidic to very basic, small to large, *etc.*) but also a single protein can change in physical character by structural rearrangement (as in denaturation). Proteins are also large enough such that physical interactions with surfaces often involve only one facet of the larger protein structure at a time. The surface(s) of the protein with which the stationary phase interacts depends on the stationary phase chemistry, but also the mobile phase conditions. For example, a protein that has both a hydrophobic region and a highly basic region can be separated by reversed phase and by cation exchange mechanisms. The behaviours in these different types of separation will be largely independent, as if they were different molecules. This heterogeneity of physical behaviour with respect to each individual interaction event can confound the development of effective methods. Even a homogenous protein of perfectly unchanging structure will chromatograph essentially as a superposition of all of the divergent facets across its surface that it presents to the stationary phase. Furthermore, the chromatographic adsorption event itself is likely to induce conformational changes.[2] The fundamentals of the chromatographic behaviour of proteins were studied primarily in the 1980s and was well summarized by Fred Regnier in 1987.[3] There continues to be studies that refine our detailed knowledge of this area using modern approaches,[4] but in many regards there remains much art in intact protein separation sciences.

General principles for protein separations:

(i) Multiple protein conformations are possible
   - *Solutions: Force the protein into a single conformation; with a complex mixture of proteins this may not be possible*

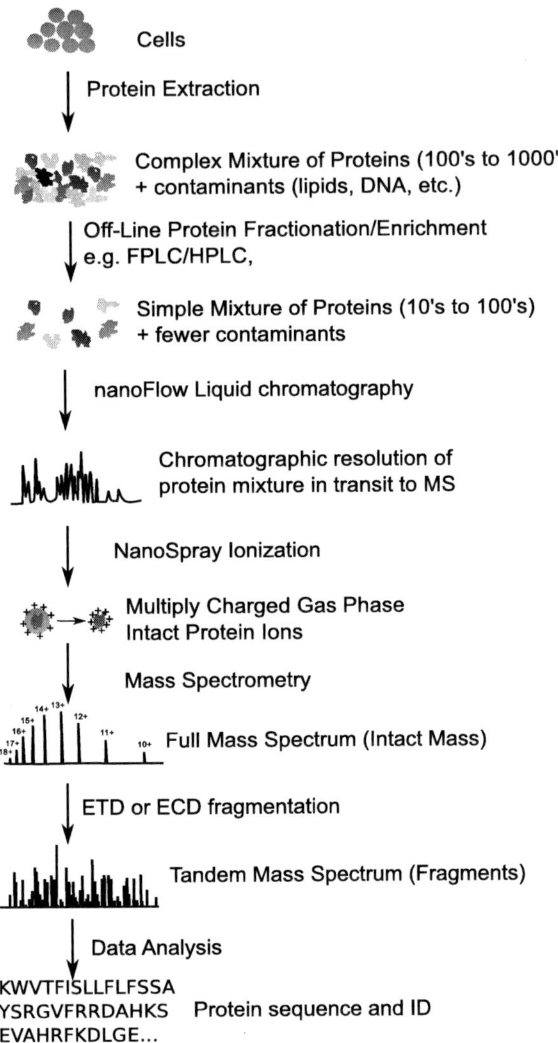

Figure 4.1 An outline of a typical intact protein LC-MS method. The early preparation step may generate several fractions to be later carried through the rest of the analysis. At each step in this process there are many approaches, each with its own unique utility. These may be combined in many different ways. Thus, there is a wide variety of intact protein LC-MS methods that have specific targets or relatively narrow focus.

ⓘ Solubility can be a major issue
- *Solutions: Choose conditions both in the mobile and stationary phases that promote solubility; again, clearly this is not possible for all proteins simultaneously*

- ⓘ Proteins exhibit poor mass transfer kinetics
  - ■ *Solutions: Raise temperature, choose wider pore or non-porous stationary phases*
- ⓘ Proteins vary greatly in character
  - ■ *Solutions: Know your protein(s)' physical characteristic or choose the physical characteristics of the proteins your wish to find. Intact protein LC-MS methods will only be effective for a subset of proteins*

4.2.1 Understanding Proteins

Effective intact protein LC-MS method development involves a good deal of knowledge and thought about protein chemistry. If knowledge of protein chemistry is weak a more thorough review of protein chemistry is recommended. For present purposes we will quickly review some of the more fundamental aspects that are relevant to LC-MS.

The most important factors that can be derived with some accuracy directly from primary sequence are size (mass), hydrophobicity and isoelectric point (pI). Even if these are not known, it is essential to think broadly in these terms. The charge of a protein can have dramatic effects on its structure, solubility, chromatography, and ionization. Thus, the pH of the chromatography must be more closely considered than in peptide and small molecule LC-MS. Table 4.1 list the pK_as of the acidic and basic amino acids as well the pK_as of the N- and C-termini. The true pI will not directly match some sum of these effects, but it is useful to keep these approximate numbers in mind.

Ultimately it is the behaviour of specific exposed facets of the protein, rather than the protein as a whole, that is generally responsible for a given interaction, so density of these groups should also be considered. For example, dense patches of the highly basic amino acids arginine and lysine often present a unique solvent-accessible patch or loop. Such effects, however, are often not clear from the amino acid sequence as many of these patches arise from

Table 4.1 The pK_as of the ionizable amino acids and the peptide termini. Many, but not all, LC-MS methods operate at low pH (<pH 4) where for the most part the acidic groups are neutral and the basic groups are positively charged

Group	pK_a
Arginine	12.5
Lysine	10.5
Tyrosine	10
Cysteine	8
Histidine	6
Glutamic acid	4
Aspartic acid	3.9
α-N-terminus	8.8–10.6
α-C-terminus	1.8–2.6

tertiary structure. It is generally a fair approximation to assume that nearly all ionizable groups and many of the hydrophilic groups are on the surface and many of the hydrophobic groups are buried in the core in a protein's native or near native form. This makes the prediction of hydrophobicity very difficult. There are several metrics of hydrophobicity that may be used to approximate the behaviour of a given peptide, but these tools are rarely useful with a protein because of the dominance of tertiary structure. Although some proteins chromatograph in a near-native state in reversed phase chromatography, hydrophobic interactions most frequently involve at least some if not extensive denaturing of the protein to expose the inner hydrophobic core.

The size of a protein is important in LC-MS in several respects, which are discussed more in sections below. In general, however, the difficulty of analysis generally scales with size. Larger proteins exhibit more variable behaviour, and poorer recovery, solubility and detection limits. These issues relate directly to the issues discussed above, as there are both more surface area and more potential conformations that the protein may take.

Another issue to consider is covalent protein modifications. These occur both *in vivo* and *in vitro*. When present, *in vivo* post-translational modifications are often a focus of the analysis and intact protein LC-MS of such closely related species presents its own challenges if more than profiling is desired. *In vitro* covalent modifications are more frequently an unintended nuisance that introduce yet another level heterogeneity to the sample. Methionine oxidation is particularly common and hard to avoid, but many other less abundant *in vitro* processing artefacts are possible. There are relatively few studies of these artefacts, and some overlap with *in vivo* non-enzymatic products. For example, oxidation processes also affect other amino acids such as lysine and arginine which typically lose mass (–1 Da and –27 Da respectively) rather than gain mass. Formylation occurs *in vivo* as a result of natural biological processes but also arises from oxidation during processing. These types of oxidation that generate aldehydes can react with almost any amine to form Schiff bases producing a near infinite set of products. Many common reagents used in processing of biological samples have also proven reactive to proteins. For example, the common serine protease inhibitor AEBSF or 4-(2-aminoethyl) benzene-sulfonyl fluoride hydrochloride, which is used to help recover intact proteins from biological samples has been shown to react covalently with analyte proteins as well as its target. Gel separation of proteins imparts various *in vitro* modifications. Electrophoresis inherently takes place in an electrolytic cell that produces oxidation and reduction products, including oxygen and other reactive oxygen species. Overall, these modifications lead to small shifts in retention time and contribute to the appearance of poor peak shape in LC-MS analyses.

4.2.2   HPLC Instrumentation

Most modern LC-MS work, intact protein or otherwise, is in a nanoflow (aka capillary) format with flow rates in the hundreds of nL/min range. Nanoflow LC columns are often made from fused silica tubing (50–150 μm internal diameter) pulled to a fine tip with a laser tip puller and the HPLC stationary phase is packed directly into this integrated column and nanospray emitter. A similar setup is also sometimes achieved by a small-format column that is coupled to a separate MS ionization source. It is not uncommon to use slightly larger format columns and use a post-column split to achieve proper flow for efficient MS ionization and for other purposes simultaneously. For example, fractions may be collected or an alternative detection technique may be used in parallel. Clearly this results in loss of sensitivity in the mass spectrometer, yet other purposes are served.

Two methods of mobile phase delivery are used today: direct flow and split flow. Direct nanoflow pumps are in common use; however, similar results may be achieved with a non-nanoflow capable pump by a simple pre-column split, controlled by the length and diameter of the tubing used on the waste leg. Samples are introduced by a low dead volume autosampler, a manual injector or by 'bomb loading' (the off-line introduction of sample from a tube on to the column by gas pressure).

4.2.3 Stationary Phase Morphology

For most intact protein LC-MS methods a pore size of at least 300 Å should be used. Pore size can have a major effect on the mass transfer kinetics. Poor mass transfer kinetics is one of the biggest challenges of intact protein LC-MS and thus proper consideration of pore size is essential. If the size of a protein is of the order of the pore size and it enters a pore it will be very slow to exit. Thus, for very large proteins an even larger pore size is essential. Even some smaller proteins may benefit from larger pores. Keep in mind that larger-pore material will have a lower surface area and lower loading capacity, *i.e.* the column will become overloaded by less material, causing degradation of chromatography. Non-porous materials have been shown to exhibit some excellent chromatography on proteins, but with a loss in loading capacity.

The particle size is another important consideration. A 5 µm particle size is common but smaller particle sizes such as 3 µm can yield better resolution at the cost of increasing back-pressure. However, this is a general effect not specific to protein LC-MS.

4.2.4 Column Temperature

Column temperature is neglected in most LC-MS analyses. For intact protein analysis the temperature can be very important in chromatographic behaviour. Temperature affects the two most critical aspects of intact protein separations: protein conformation and mass transfer kinetics. The maintenance of a single protein conformation is usually preferable (unless the conformations themselves are being studied). Of course an ensemble of proteins is never in a truly singular homogeneous conformation. Thus, the real solution is the rapid interconversion between very similar conformations. Temperature can address both of these by bringing the proteins into a denatured state and then speeding up the kinetics of interconversion. This also simultaneously solves the problem of mass transfer kinetics between the mobile phase and stationary phase. The alternative approach is to deliberately maintain a lower temperature and other conditions where the native structure is maintained. This simpler solution can work well for some proteins.

4.2.5   Mobile Phase Composition

The use of organic solvents in intact protein LC-MS is common. These are the same as found in all types of HPLC and LC-MS, primarily acetonitrile and methanol. It is important to remember, however, that the solvent has a major effect on the protein as well. Any substantial amount of organic solvent usually rapidly denatures proteins, and may cause solubility issues and possibly bulk protein precipitation. Nonetheless, their use is often an essential component of the chromatography.

It is generally a good idea to buffer mobile phases. This will provide more reproducible results, and buffers and mobile phase additives can be a powerful tool in manipulating the chromatography and the behaviour of the protein. Some of the common volatile acids and buffers used in LC-MS are listed in Table 4.2.

4.2.6   Matrix Effects

Most intact protein LC-MS methods start with an at least somewhat crude mixture of proteins and sometimes even lightly processed biological fluids. Additionally, other non-protein components may be present. All of these components, less the species being analysed, are collectively referred to as the matrix. The complexity of the matrix is of great importance as it affects column loading capacity and increases the likelihood of coelution of components.

Table 4.2   The pK_as of the common volatile mobile-phase additives used for LC-MS methods and their buffer ranges. The pH used and the ionic strength/concentration used can have dramatic effects on both the chromatography and the ionization. These effects are often contradictory and require a balance

Additive or buffer	pK_a	Buffer range
TFA[a]	0.5	—
Formic acid	3.8	—
Ammonium formate	3.8	2.8–4.8
Acetic acid	4.8	—
Ammonium acetate	4.8	3.8–5.8
Ammonium bicarbonate	6.3/9.2/10.3	6.8–11.3
Ammonium acetate	9.2	8.2–10.2
Ammonium formate	9.2	8.2–10.2
Triethylamine acetate	11.0	10.0–12.0

[a]TFA causes severe ionization suppression during electrospray ionization and should be avoided.

Even if interfering components are of different masses they may affect ionization and the ability to isolate and enrich the analyte in the gas phase (see also Chapter 2). The effect of the matrix on any given analyte can shift retention times, degrade the quality of separation and affect ionization efficiencies and thus quantification. It is important to maintain an awareness of this and when possible account for these effects. The simplest approach is to maintain the same matrix, or a surrogate thereof, throughout the work. One of the distinct advantages of intact protein LC-MS rather than MS alone is that the ionization efficiency effects of the matrix are significantly reduced. The physical separation of the liquid chromatography results in a dramatically less complex mixture at the point of ionization, thus increasing ionization efficiency and dynamic range. Unique to intact protein LC-MS, protein–protein interactions can cause matrix effects of high specificity where the matrix effect on one protein may be dramatic. This is sometimes used to study such interactions (see below).

4.2.7 Sample Preparation

Sample preparation for intact protein LC-MS is a much larger topic than can be addressed here; however, there are aspects specific to LC-MS that many experts in protein purification and manipulation are typically unaware of. These issues are also often the first cause of failure on a project. Most protein manipulations are performed in non-volatile buffers, such as phosphate-buffered saline (PBS), Tris-buffered saline (TBS) or Tris, at relatively high molarity. This is a poor place to start, since additional steps will be required, such as a buffer exchange or solid-phase extraction step, further compounding losses. This is not as relevant to bottom up methods, as the digestion process also serves as a buffer exchange step. In intact protein LC-MS the protein or protein mixture is often injected without further processing beyond pre-fractionation. Thus, it is important to design protein processing steps to bring the sample into a buffer system compatible with the LC-MS analysis, often by slightly modifying standard molecular biology protocols. Another option is that if the buffer system is compatible with the chromatography but simply not with the mass spectrometer, a buffer exchange can be performed on the analytical column by loading off-line, washing with the LC-MS starting buffer and then running on-line LC-MS.

4.2.7.1 *Prefractionation*

Often some level of off-line prefractionation is performed on samples to further reduce the complexity of the sample for purely analytical purposes beyond those required by the biological question at hand. There is much research into the most effective, reproducible and orthogonal means of reducing the complexity of complex intact protein mixtures before on-line LC-MS analysis.[5] Gel electrophoresis techniques provide an excellent protein-level separation that has been used for many years; however, they have many disadvantages, including poor recovery of the protein from the gel and the introduction of covalent artefacts. Some of these approaches include electrophoretic techniques that function in solution rather than in a gel, such as field flow fractionation,[6] the commercially available Gelfree system from Protein Discovery[7] or the Offgel system from Aligent Technologies.[8,9] Other approaches simply use liquid chromatography orthogonal to the primary analytical chromatography. For example, ion exchange techniques (as describe below) are often used as a prefactionation step before reversed-phase LC-MS. Many of these pre-fractionations are identical or similar to techniques used for years in protein purification, as can be seen in Chapter 3.

4.2.7.2 Avoiding Losses

Another issue in sample preparation for intact protein LC-MS is loss of protein to surfaces. Most workers in the field use polypropylene autosampler vials because of the relatively lower protein adsorption to the surface. Glass vials should be avoided. There are chemically derivatized glass vials (*e.g.* silanized vials) that reduce the effects, notably ion pairing with silanols, which result in high protein adsorption. If a particular protein or class of proteins is being studied, these effects can be considered specifically. These effects are also present in any upstream manipulations where years of practice generally have resulted in near-exclusive use of polypropylene by molecular biologists (Ependorf tubes, Falcon tubes *etc.*). This use of polypropylene is usually carried forward without thought by the non-expert. Downstream within the LC-MS system the tubing chosen for plumbing and for nano-LC columns can also be an issue. Many modern LC-MS systems use fused silica tubing that in theory should suffer from the same issues as glass autosampler vials. In reality this effect is most often quickly obliterated by saturation of the surface with adsorbed protein, passivating the surface. Considering this, it is advisable to run non-precious standards on new systems and tubing. Nano-LC columns are often 'blocked' after they are made by running a cheap standard peptide or protein to passivate sites of irreversible binding and reduce future losses.

4.2.7.3 Loading Buffer

In most HPLC and LC-MS methods it is considered best practice for the sample to be injected in the solvent starting conditions. This is also generally true for intact protein work, but there are some exceptions worth considering. Although it is far from ideal, sometimes using a different solvent/buffer system can enable temporary solubility for injection purposes. Precipitating the sample in the LC system should be avoided, but does not always result in catastrophe and may ultimately be necessary. A related instance where this may be necessary is manipulation of the conformational state before starting the analysis. For example, introducing large amounts of acid to denature proteins can help equilibrate them into a form that will chromatograph well without significant effect on the solvent strength of the loading buffer, at least for reversed-phase methods. Such approaches must not interfere with the chromatography.

For example, use of higher concentrations of organic solvent in the injection buffer in reversed-phase analyses is a very bad idea since it will introduce a slug of strong solvent at the beginning of the analysis.

4.2.8 Choice of Stationary-Phase Chemistry

The chemistry of the stationary phase used in an LC-MS separation dominates the nature and properties of the method. Intact protein LC-MS presents a greater challenge than small-molecule or peptide LC-MS such that careful consideration of the mode and mechanism of chromatographic separation is essential to effective method development.

4.2.8.1 Reversed-Phase Liquid Chromatography

Reversed-phase liquid chromatography (RP-LC) is nearly synonymous with HPLC and LC-MS in the minds of many; however, in intact protein analysis it suffers from many potential problems. The typical reversed-phase conditions of low pH and high and changing organic solvent concentrations are ideal conditions to denature proteins. Typically proteins are in at least a partially denatured state as they traverse a reversed-phase column, but multiple or changing conformations can be problematic. The problems most typically observed are peak broadening (due to conformational heterogeneity and poor kinetics), peak asymmetry (likely due to a more extensive conformational heterogeneity approaching distinct populations), multiple peaks for a pure protein (truly distinct conformational populations that interconvert on a longer time scale than the chromatography), poor recovery (irreversible binding), and high on-column carry-over (conformational changes allowing re-entry into the mobile phase on later gradients). These problems make RP-LC far from easy and universal in protein LC-MS methods. For these reasons, careful selection of the column chemistry is important. There are several basic reversed-phase chemistries and many subtle differences between columns and manufacturers. We present here the most basic selection of phases as an introduction to possible phases. The chemical structures of theses stationary phases are shown in Figure 4.2.

- ① *ODS (C18):* This common choice for peptide and small-molecule drug analysis is rarely an excellent choice for protein separations. Nonetheless, its use is not uncommon
- ① *Octyl (C8):* This shorter chain is a substantially better choice and popular for intact protein LC-MS. The shallower bed of hydrophobic stationary phase reduces many of the problems detailed above, as the protein cannot enter the stationary phase as deeply and will thus not induce as much conformational change
- ① *Butyl (C4):* This is a common reversed-phase chemistry for protein LC-MS. It has even better characteristics than C8 regarding how deeply the protein can enter the phase and is essentially self end-capped

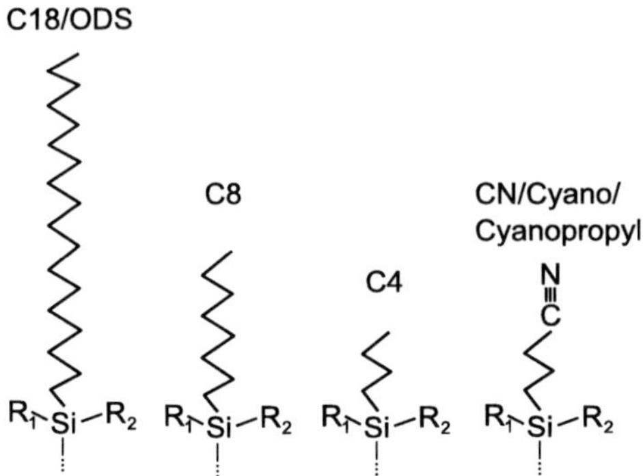

Figure 4.2 Examples of common reverse-phase chemistries. The reaction used to generate silica-based reverse-phase stationary phases gives control over three out of four silica bonds. The R_1 and R_2 ligands above are most commonly methyl groups, but other groups are sometimes used in order to disrupt phase collapse.

(i) *Cyano (CN):* This is an underappreciated and highly versatile stationary phase. With chemistry remarkably similar to acetonitrile, it is a relatively weak hydrophobic phase and can require less organic to elute proteins. It also has even better kinetics and end-capping. Interestingly, it may also be used in a normal-phase mode for many proteins

4.2.8.1.1 Subtleties of the Chemistry of Reversed Phase Columns. Most reversed-phase columns consist of the namesake hydrocarbon chain covalently attached to a silica atom on the particle surface. Typically there are, due to the synthesis, two methyl groups also bound to the same silica atom, but sometimes other groups are used. Some 'AQ' (aqueous) columns use these sites to prevent phase collapse by introduction of a moderately hydrophilic group. Also, the underlying silica can contribute a significant effect to the chromatography. Bare silica is generally normal-phased in nature but also has cation exchange character through the presence of negatively charged silanols. This is a special concern for very basic proteins. Strong ion pairing agents such as trifluoroacetic acid (TFA) are used in off-line separations to mitigate unwanted cation exchange effects. TFA removes these interactions in two ways: (1) it neutralizes the charge on the stationary phase by protonating negative silanols; (2) it neutralizes some of the charge on the protein by forming strong ion pairs with positive residues. In intact protein LC-MS TFA is a poor choice because of its very negative effects on electrospray ionization efficiency. The major solution to this is the removal of residual silanols by a process of end-capping. When the primary chemistry being imparted to the column is derivatized to the surface, not all sites are reacted due to steric effects of the stationary phase already present. By taking the stationary phase through a second round of derivatization with a smaller reagent a greater number of these sites may be reacted. The end-capping process thus introduces a chemistry that is an imperfect but much improved match to the primary stationary-phase chemistry. Thus, careful consideration of the presence and extensiveness of end-capping of the stationary phase can be essential. For these reasons monolithic columns that do not use silica particles, such as polystyrene divinylbenzene-based columns, have been of great interest. The overall performance of such columns, however, does not currently approach what is possible with the traditional packed beds.

4.2.8.1.2 Gradient Design. The design of the gradient for intact protein RP-LC-MS methods needs more consideration than the typical LC-MS method. Often RP-LC-MS methods are designed with very sharp gradients in order to obtain sharp peaks and improve method sensitivity. With proteins this approach can backfire. The slow kinetics and the potential for conformational change can result in worse peak profiles, even worse recovery, and carry-over. One interesting manifestation of an overly sharp gradient is the appearance of a second peak on the reverse gradient at approximately the %B associated with elution during the forward gradient. This effect results from moving through the ideal elution conditions faster than the protein can elute. This will also result in carry-over peaks. Ultimately there is an optimal gradient slope that yields the best result, and steeper is not always better.

4.2.8.2 Ion Exchange Chromatography

Ion exchange chromatography (IEC) can also be used in online LC-MS methods of intact proteins. Most off-line IEC-HPLC methods use a salt gradient to elute proteins from ion exchange columns. Similar approaches may be adopted in LC-MS by use of the volatile salts listed in Table 4.2. Although none of these are ideal salts for IEC, they nevertheless function in the same manner. High concentrations of even these volatile salts may cause problems in electrospray ionization where they can reduce ionization efficiency and slowly contaminate the source. An alternative approach is to use a pH gradient instead of ionic strength to elute proteins.[10] This technique, sometimes called *chromatofocusing*, is performed on weak cation exchange (WCX) or weak anion exchange (WAX) resins, where the ion exchange resin itself is neutralized. In general IEC techniques exhibit high recovery of intact proteins from the column and excellent selectivity and can be performed under fairly non-denaturing conditions. The charge or isoelectric point of the protein(s) being analyzed is the essential physical characteristic being leveraged in such analyses. In Figure 4.3 some common ion exchange chemistries are shown. Further materials are shown in Table 4.1.

4.2.8.3 Hydrophilic Interaction Liquid Chromatography

Hydrophilic interaction liquid chromatography (HILIC) is essentially a form of normal-phase chromatography.[11] HILIC resins consist of a hydrophilic molecule bonded to an underlying particle. More traditional normal-phase resins are generally bare particles that have innate hydrophilic character.

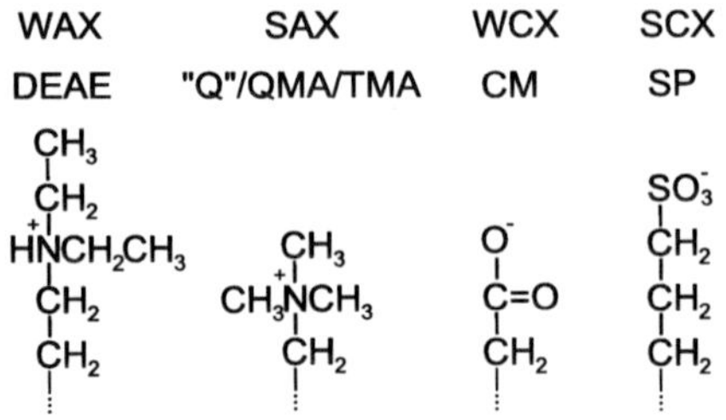

WAX	SAX	WCX	SCX
DEAE	"Q"/QMA/TMA	CM	SP

$$
\begin{array}{cccc}
\text{WAX} & \text{SAX} & \text{WCX} & \text{SCX} \\
\text{DEAE} & \text{"Q"/QMA/TMA} & \text{CM} & \text{SP} \\
\end{array}
$$

Figure 4.3 Examples of common ion exchange chemistries. There are many variations on these themes with varying substituents and branching structures. The weak-strong distinction is related to weak *versus* strong acids/bases, *i.e.* if the ion can be neturalized by acid or base. For example quaternary amines are a permanently fixed charge that cannot be neutralized by acid-base chemistry.
CM, carboxymethyl; DEAE, diethyl amino ethyl; Q, quaternary amine; QMA, quaternary methyl amine; SAX, strong anion exchange; SCX, strong cation exchange; SP, sulfopropyl; TMA, trimethylamine; WAX, weak anion exchange; WCX, weak cation exchange.

The bonded phase of HILIC typically provides better reproducibility and chromatographic resolution. The buffer system is generally LC-MS compatible, starting at high organic to low organic/high water. HILIC is often found in a mixed mode, as HILIC mechanisms can be induced by many functional groups. One example is the use of what are essentially ion exchange resins in a HILIC mode. Charged functional groups are essentially an extreme of hydrophilicity, and interactions can occur due to non-ionic effects. By running a high-to-low organic gradient these interactions may also be utilized for greater selectivity.

4.2.8.4 Size Exclusion Chromatography

Size exclusion chromatography (SEC) is a technique that separates proteins on the basis of their hydrodynamic radius. This makes it relatively redundant with MS, yet it is easily made compatible with LC-MS and there are several ways in which SEC LC-MS is useful. The most notable use of SEC LC-MS is in the study of non-covalent protein interactions and complexes. The hydrodynamic radius of the protein complex may be measured under non-denaturing and various degrees of denaturing conditions. The proteins will traverse the column together if in a complex. Similarly, different conformations of the same protein may be studied.

4.2.9 Two-Dimensional Liquid Chromatography

All of these various chromatographic techniques may be used in series online in a process termed two-dimensional liquid chromatography (2D-LC). This simply means that the sample is injected onto one column and fractions from one column are directly introduced into a second column, usually by some sort of step gradient.[12] The eluent of the first column must not interfere with retention of the sample on the second column. This is distinct from off-line prefractionation, which is considered a sample preparation step, in that it improves throughput and recovery and is often more automated, but it is not substantially different.

4.3 Mass Spectrometry

The ionization of large proteins was solved in the late 1980s with the development of electrospray ionization (ESI) and matrix-assisted laser desorption ionization (MALDI) techniques. Subsequently the major improvements in the mass spectrometric analysis of intact proteins have been better, higher-performance detection and the development of effective fragmentation/sequencing techniques. The improvements in detection have mostly come naturally with the general field of MS and proteomics, but some are specific to these more massive analytes. Even with these improvements, instruments may still need to be optimized for larger masses. The greater inertia and longer flight time of intact proteins can cause issues in the electrodynamics of the mass spectrometer. For example, trapping instruments are often coordinated in time to an ion packet; however, large proteins will arrive very late and the timing may need to be adjusted to trap these ions effectively. These larger masses can also be problematic in radiofrequency trapping because the resonant frequency is much lower. The frequency of quadrupoles, hexapoles, octapoles, *etc.* may need to be changed to account for this, at the cost of losing smaller ions. These instrument design parameters are usually not user controllable, but purchasing decisions are. The best, most versatile instruments for intact protein analysis are the high-end instruments such as Fourier transform ion cyclotron resonance (FTICR) mass spectrometers, but certain applications can use relatively inexpensive instruments, even a cheap single quadrupole. Further mass spectrometric discussion can be found in Chapter 2. Intact protein analysis performance can vary widely between similarly designed and priced instruments as a result of optimizations for this relatively less common application.

Proteins, being large, complex, and flexible molecules, are prone to form adducts—non-covalent complexes with small ions (*e.g.* sodium) or molecules (*e.g.* water)—in the ionization process. Small molecules and peptides also form adducts, but less readily. Different proteins also vary in their propensity to form adducts. This can be a problem, in that signal is diluted and the accurate intact mass of the protein can be obfuscated. In some cases adducts can interfere with detection of coeluting proteins and effect fragmentation. Adducts, by definition, coelute with the purely protonated species. They are products of the ionization process and do not reflect any differences in liquid phase behaviour.

4.3.1 LC-MS Profiling/Quantification

Many intact protein LC-MS methods are profiling methods that are limited in scope. It is not uncommon for only the masses of the proteins to be determined. Such simple approaches are useful in profiling slight modifications to or relative amounts of relatively well-characterized proteins or protein mixtures. If quantitative results are desired, it is important to consider that the intact protein LC-MS signal is actually distributed among multiple charge states. It is considered best practice to quantitate against the sum of the charge state intensities, since slight changes in ionization conditions can shift the charge state distribution without necessarily affecting the overall signal. This has the added advantage of simultaneously averaging signal, improving signal to noise. Profiling and quantification are only some of several types of useful information possible from intact protein LC-MS without full characterization *via* top down mass spectrometric sequencing.

4.3.2 Conformational Analysis and Protein–Protein Interactions

There are several contributions to measuring and understanding conformation and non-covalent protein–protein interactions from the MS side of the LC-MS technique. In addition to serving as a mass detector for chromatographic techniques that contribute conformational and interaction information, the conformations and interactions themselves can be addressed in the gas phase. Solution phase complexes and conformations may be transferred relatively unperturbed, *via* nanospray ionization for example, to the mass spectrometer where they can be interrogated in several ways.[13]

Ion mobility spectrometry-mass spectrometry (IMS-MS) is a technique that first measures the hydrodynamic radius ions *via* an ion mobility separation before mass spectrometric analysis. Ion mobility involves the gas phase separation of ions in a pressure regime (near atmospheric pressure) where interactions with an inert bath gas dominate. The energy imparted in these collisions is low and generally not dissociating, nor do they strongly affect conformation. The frequency of these collisions and the degree to which they retard motion is a function of the average cross-sectional area of the tumbling protein. The charge state of the protein will affect the force imparted by the electric field. In this way the flight time is used to determine the ratio of charge to size.[14] Once the charge state is determined *via* mass spectrometry, the size is easily extracted. Ion mobility measurements are much faster than liquid chromatography and slower than mass spectrometry, such that it is effective as another level of separation between the two. In this way conformational parameters may be directly determined within an LC-IMS-MS analysis.

Gas phase complexes may also be studied by fragmentation. Techniques such as high-energy collision induced dissociation (HCD) or surface induced dissociation (SID) are capable of disrupting non-covalent complexes without fragmenting covalent bonds. This can then be used to distinguish between non-covalent and covalent bonds (it is or is not a complex) and determine the relative composition and component identity of these complexes during a non-denaturing intact mass LC-MS analysis.

4.3.3 Top Down Sequencing

In the 2000s the capacity to sequence whole proteins was developed, with first electron capture dissociation and then electron transfer dissociation demonstrating near-complete sequence coverage on increasingly larger intact proteins.[15,16] Much of this work, however, is with statically infused samples rather than with on-line LC-MS. The speed and efficiency of the fragmentation process continues to improve and is currently quite capable of chromatographic timescale fragmentation of reasonable-sized proteins.[17] Figure 4.4

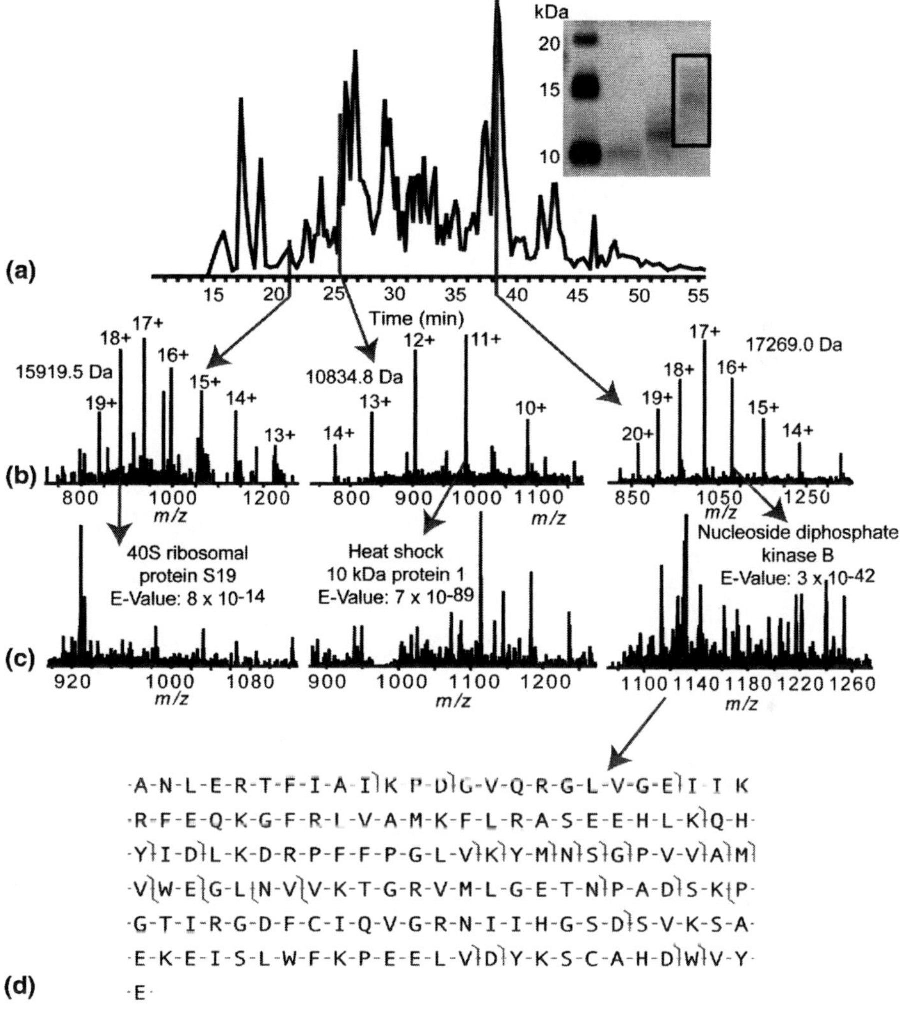

Figure 4.4 A figure from Lee *et al.*[18] Examples selected from an LC-MS/MS injection of a fraction from a Tris–glycine GE run. (a) A base-peak chromatogram, with (b) charge state distributions from selected retention times. (c) Abundant charge states (above the arrows) were targeted for fragmentation. Fragmentation mass spectra for each protein are shown along with the corresponding identifications and E-values. A fragmentation map (d) results from the matching fragment ions of nucleoside diphosphate kinase B. The protein is N-terminally acetylated. Reprinted with permission from *J. Am. Soc. Mass Spectrom.*, 2009; **20**(12), 2183. Copyright 2009 Journal of the American Society for Mass Spectrometry.

shows such an online top down LC-MS analysis from the group of Neil Kelleher, a major advocate of such approaches.[18] Further principles of top down analysis and bottom up analysis can be found in Chapter 1.

Both ECD and ETD function by imparting a single low-energy electron into a molecular orbital of the protein. This results in a very rapid dissociation of the protein such that the protein is fragmented at a near-random location along the protein backbone at the N(H)-C(R) bond, within an amino acid residue rather than between residues. The N-terminal side of proline is excluded from this process as it contains a ring structure that spans this bond. The resulting fragments are termed c-ions and z-ions. The mass spectrometry parameters for top down fragmentation are more variable, primarily due to the innate variety of proteins, than for relatively similar small peptides that fragment in a relatively narrow energy range. For ECD/ETD top down analysis of intact proteins short reaction times are used (1–10 ms), and good if not complete sequence coverage can be achieved across a good number of proteins. The complementary technique of proton transfer reaction (PTR) is a useful means of reducing the charge of highly charged fragments such that they are more easily detected.

It should be noted that there is an inherent rather than purely technological challenge in the fragmentation of large proteins. The larger the protein, the more the starting parent ion signal will be diluted into a larger number of fragment ions. A six amino acid peptide will fundamentally yield about a ten-fold decrease in signal from the intact precursor ion signal, before other losses, simply because one ion becomes ten ions. Similarly, a 101 amino acid protein automatically has fragment ions about 0.5% of the precursor parent ion signal.

Collision-induced dissociation (CID) and high-energy collision dissociation (HCD) are also capable of top down analysis on smaller proteins. This is particularly useful on low-resolution instruments, which less commonly have ECD or ETD and anyway cannot resolve meaningful information from the fragmentation spectra of larger proteins.

True on-line top down intact protein LC-MS characterization requires a high-resolution mass spectrometer for proteins greater than about 25 kDa (*i.e.* most proteins). The number of fragment peaks produced is great enough such that the density of peaks would result in extensive interferences and loss of information without high resolution. Such high resolution data is also extremely useful in correct identification of fragment peaks. Small proteins may be analysed *via* top down analysis on low-resolution mass spectrometers, such as ion traps, with some challenge. There are currently two mass analysers that can provide such data: Fourier transform ion cyclotron resonance (FTICR) and the Orbitrap mass analyser. FTICR is the more capable instrument for top down analysis. The Orbitrap has many advantages over FTICR but suffers from the distinct disadvantage of difficulty in measuring large masses. Thus, acquiring an intact mass of the precursor holoprotein of substantial size in an orbitrap can be difficult. This does not, however, limit its capacity to detect fragment ions. Top down data can also be challenging to interpret. Widely available software is often only a starting point that must be followed up with either manual confirmation or more sophisticated in-house software analysis.

4.4 Notes

The LC-MS analysis of intact proteins is a challenging yet powerful approach in proteomics. It continues to be further developed and refined for a variety of purposes, and even relatively simple intact protein LC-MS methods can provide necessary complementary information to the more ubiquitous bottom up or peptide-level MS analysis. The many challenges faced in the development of online LC-MS methods for intact protein analysis are best addressed through an awareness of the fundamentals of chromatography and protein behaviour. The limitations and capacity of the mass spectrometry also need to be considered, but this side of such methods frequently suffers from fewer subtleties.

References

1. R. L. Cunico, T. Wehr and K. M. Gooding, *Basic HPLC and CE of Biomolecules*, 1st edn., Bay Bioanalytical Laboratory, Richmond, CA, 1998.
2. X. M. Lu, K. Benedek and B. L. Karger, *J. Chromatogr.*, 1986, **359**, 19.
3. F. E. Regnier, *Science (New York, N.Y.)*, 1987, **238**, 319.
4. M. F. Engel, A. J. Visser and C. P. van Mierlo, *Proc. Natl. Acad. Sci. U S A*, 2004, **101**, 11316.
5. Y. Fang, D. P. Robinson and L. J. Foster, *J. Proteome Res.*, **9**, 1902.
6. P. Reschiglian and M. H. Moon, *J. Proteomics*, 2008, **71**, 265.
7. J. C. Tran and A. A. Doucette, *Anal. Chem.*, 2009, **81**, 6201.
8. B. Manadas, J. A. English, K. J. Wynne, D. R. Cotter and M. J. Dunn, *Proteomics*, 2009, **9**, 5194.
9. S. Elschenbroich, V. Ignatchenko, P. Sharma, G. Schmitt-Ulms, A. O. Gramolini and T. Kislinger, *J. Proteome Res.*, 2009, **8**, 4860.
10. N. L. Young, P. A. DiMaggio, M. D. Plazas-Mayorca, R. C. Baliban, C. A. Floudas and B. A. Garcia, *Mol. Cell. Proteomics*, 2009, **8**, 2266.
11. A. J. Alpert, *J. Chromatogr.*, 1990, **499**, 177–196.
12. Z. Tian, R. Zhao, N. Tolic, R. J. Moore, D. J. Stenoien, E. W. Robinson, R. D. Smith and L. Pasa-Tolic, *Proteomics*, 2010, **10**, 3610.
13. M. Zhou and C. V. Robinson, *Trends Biochem. Sci.*, 2010, **35**, 522–529.
14. E. Jurneczko and P. E. Barran, *Analyst*, 2010.
15. R. A. Zubarev, N. L. Kelleher and F. W. McLafferty, *J. Am. Chem. Soc.*, 1998, **120**, 3265.
16. J. E. Syka, J. J. Coon, M. J. Schroeder, J. Shabanowitz and D. F. Hunt, *Proc. Natl. Acad. Sci. U S A*, 2004, **101**, 9528.
17. A. Chi, D. L. Bai, L. Y. Geer, J. Shabanowitz and D. F. Hunt, *Int. J. Mass Spectrom.*, 2007, **259**, 197.
18. J. E. Lee, J. F. Kellie, J. C. Tran, J. D. Tipton, A. D. Catherman, H. M. Thomas, D. R. Ahlf, K. R. Durbin, A. Vellaichamy, I. Ntai, A. G. Marshall and N. L. Kelleher, *J. Am. Soc. Mass Spectrom.*, 2009, **20**, 2183.

LC-MS(/MS) of Trypsin-Hydrolysed Proteins

SERONEI C. CHEISON[1,2] AND ULRICH M. KULOZIK[3]

[1] Zentralinstitut für Ernährungs- und Lebensmittelforschung (ZIEL):
Bioactive Peptides and Protein Technology, Technische Universität
München, Weihenstephaner Berg 1, D-85354 Freising, Germany; [2] School of
Public Health and Community Development, Maseno University, Private
Bag, Kisumu, Kenya; [3] Zentralinstitut für Ernährungs- und
Lebensmittelforschung (ZIEL) Abteilung Technologie, Lehrstul für
Lebensmittelverfahrenstechnik und Molkereitechnologie, Technische
Universität München, Weihenstephaner Berg 1, D-85354 Freising, Germany

5.1 Introduction

Proteins are hydrolysed with either enzymes or chemicals (acid or alkali).
Enzymes are preferred because their hydrolysates are produced under mild
temperatures and pH, with the products being fairly homogeneous and predictable if an enzyme with known hydrolytic patterns is used. Chemical
hydrolysis, on the other hand, releases heterogeneous products with the likelihood of foulant products like lysino-alanine. During hydrolysis, upon the
cleavage of a peptide bond with the addition of a water molecule, the resulting
peptides and/or amino acids each possesses an amino (N-) or carboxy (C-)
terminal (Figure 5.1).

The terminal groups are ionized depending on the hydrolysis pH and temperature (Figure 5.2).[1] Because the carboxyl group is deprotonated at high pH,
the free hydrogen ions (H^+) lead to a drop in the pH. Likewise, when hydrolysis

RSC Chromatography Monographs No. 15
Protein and Peptide Analysis by LC-MS: Experimental Strategies
Edited by Thomas Letzel

Published by the Royal Society of Chemistry, www.rsc.org

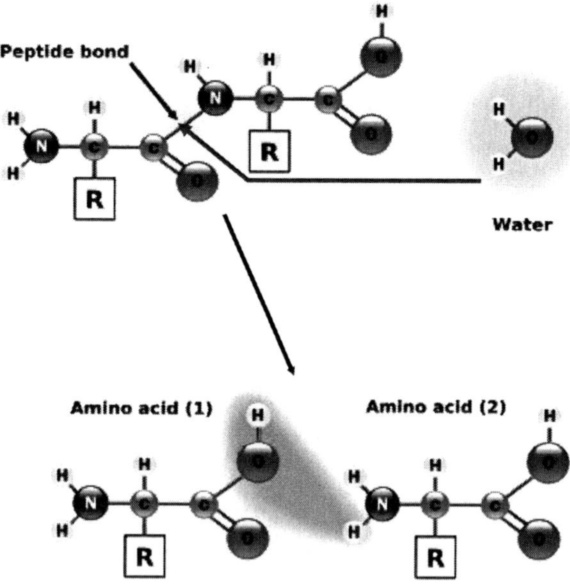

Figure 5.1   Typical hydrolysis process showing the addition of a water molecule during peptide bond breakdown.

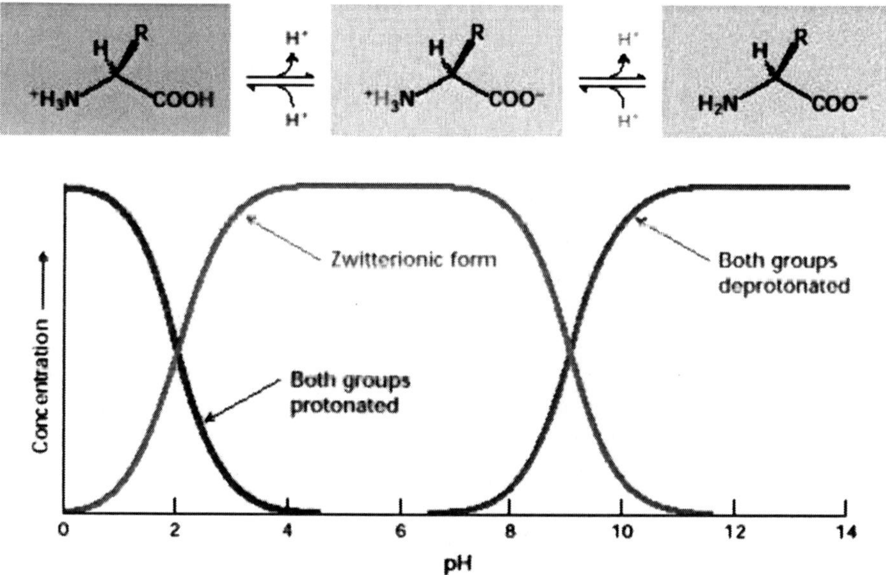

Figure 5.2   Influence of the pH on the ionization of an amino acid or peptide N- and C-terminal regions. During hydrolysis at pH > 6, the pH generally drops and is kept constant using an alkali whose volume is proportional to the number of peptide bonds cleaved. Hydrolysis at low pH leads to a rise in the pH which is adjusted using an acid.

is carried out at low pH the pH increases. This change in the pH during hydrolysis is harmful to the stability of the enzyme as well as its activity. To keep the pH constant, alkali (usually NaOH) and acid (usually HCl) is added to adjust the pH, a process which may be managed manually by continuous addition of the pH-correcting solutions. There are autotitrators available on the market, which regulate the pH to narrow limits based on preset conditions. Autotitrators are managed using personal computers or in-built software (Figure 5.3). In either case, the amounts of the pH-correcting solutions are recorded and used to calculate the degree of hydrolysis (DH) according to the pH-stat method. The DH can be understood simply as a measure of the extent to which a protein is hydrolysed. Thus, for a protein that is not yet hydrolysed at all the DH equals 0%, whereas a DH of 100% implies that a protein is completely hydrolysed to free amino acids. In practise, during enzyme

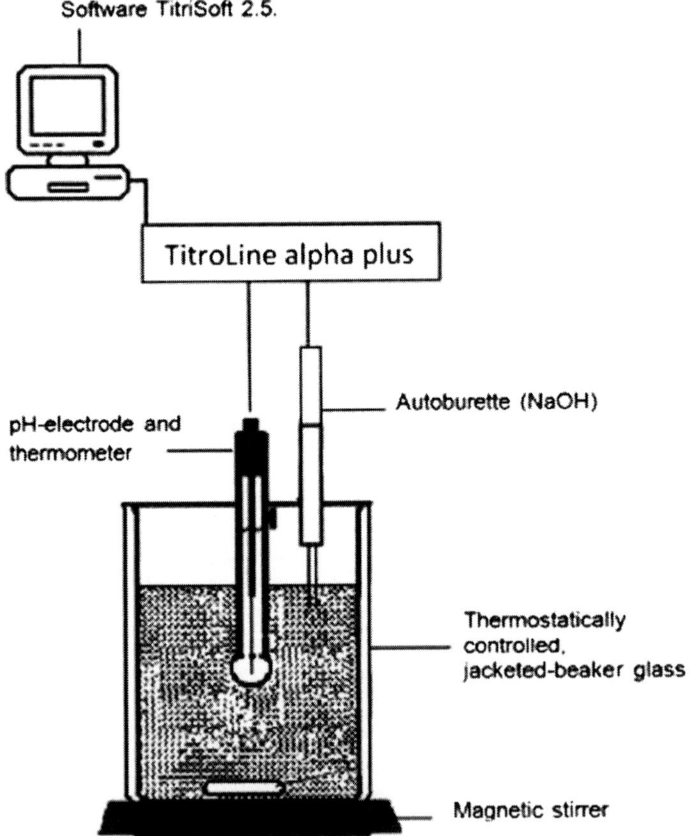

Figure 5.3 A schematic presentation of a typical enzyme hydrolysis setup with an autotitrator managed using a personal computer software. The reactor is a jacketed beaker provided with a thermostated control for temperature regulation. The pH is continually monitored and adjusted using an acid or an alkali.

hydrolysis, DH values in the range of 10–20% are achievable. The DH relates quite well with peptide bitterness, which increases with increasing DH due to the release of short peptides,[2] allergenicity and immunogenicity,[3] peptide bioactivity[4] and the functional properties of hydrolysates.[5,6]

The assembly shown in Figure 5.3 has provisions for temperature control using a thermostatically controlled water bath and a pH adjustment run using an autotitrator.[7] To monitor protein hydrolysis, the DH is used. By definition, the DH is a ratio of the total peptide bonds in a proteins hydrolysed (Equation 5.1):

$$\text{DH} = h/h_{\text{tot}} \tag{5.1}$$

where h is the number of peptide bonds hydrolysed and h_{tot} is the total number of peptide bonds in a protein. In practice, hydrolysis does not involve a single molecule of the protein, meaning a more meaningful interpretation of the quantity h_{tot} is therefore a weighted sum of the available peptide bonds in a proteins based on molar concentrations. To determine the DH during hydrolysis, proportionality between the volume of the pH-correcting solution is used when the DH is determined according to the pH-stat method.[8] The pH-stat method (pH is kept *static* or constant), the volume and concentration of the alkali or acid is taken into consideration according to Equation 5.2:

$$\text{DH} = V \times N \times (1/\alpha) \times (1/M_{\text{p}}) \times (1/h_{\text{tot}}) \times 100 \tag{5.2}$$

where V is the alkali/acid consumption in mL, N is the normality of the alkali/acid, α is the average degree of dissociation of the α-NH/COOH groups, M_{p} is the weight of protein (in grams) and h_{tot} is the total number of peptide bonds in the protein substrate (calculated to be 7.2 meq g^{-1} protein for β-lactoglobulin (β-Lg)[7] and 8.8 meq g^{-1} for whey proteins).[1]

Other methods used to quantify the DH are osmometry, reaction of the liberated NH_2 groups with a chromophore-forming agent such as ninhydrin[9] and trinitrobenzenesulphonic acid (TNBS). These other methods are reviewed by Cheison *et al.*[10] and Adler-Nissen.[11] The pH-stat method is the most popular and straightforward and relates linearly to the other methods.[12,13]

Protein hydrolysis is followed by mass spectrometry in order to elucidate the amino acid sequences of resulting peptides. Usually, a protein is hydrolysed using a protease with known specificity for peptide bonds, like trypsin (EC 3.4.21.4), which breaks down peptide bonds on the C-terminals of arginine and lysine.[14] Subsequently, the peptides are then analysed using various MS protocols (see Chapter 1) following the bottom up protocol.[15] In the food industry, with the emergence of interest in bioactive peptide research,[16] it has become increasingly necessary to involve MS.[17]

Successful MS of peptides depends on the transfer of ionized peptides into gaseous phase, with minimum destruction of the analyte, followed by the separation of the ions and finally their detection. Several ionization methods are frequently used for the analysis of protein hydrolysates. Among these are methods based on atmospheric pressure ionization (API)[18] and matrix assisted laser/desorption ionization (MALDI).[19,20] After ionization of the analyte, the ions can be detected by the mass analyser, *e.g.* the time-of-flight (TOF) or the quadrupole analyser or ion-trap. A combination of ionization source and mass analyser is a *mass spectrometer*.

Two of the most popular mass spectrometers are the Nobel Prize-winning inventions ESI-TOF-MS[18] and MALDI-TOF-MS.[19] These two methods differ basically in the manner of ion generation, being based on ionization under pressurized gas (usually nitrogen) (ESI) and ionization using laser power and a suitable matrix material (MALDI).

Hydrolysis may involve mixed proteins like whey protein concentrate (WPC) or isolate (WPI). However, identifying the source of peptides in such mixed proteins, especially where amino acid sequences are similar at some regions in the protein, is a difficult task. Our work uses purified bovine β-Lg, the major whey protein of ruminant species (~ 3 g L^{-1}), which is also present in the milk of some other species, but not in human milk. Therefore β-Lg is one of the main causes of cows' milk allergy in humans, especially in infant formula. It belongs to the protein family lipocalins. Bovine β-Lg is a small protein with 162 residues and molecular weight 18.4 kDa. Several variants of β-Lg have been identified, even within one single species. For example, there are at least nine variants from the cow (*Bos taurus*), which are labelled as A, B, C, D, E, H, I, J and W. The three common variants are A, B and C, which respond differently to heat.[21] β-Lg has five cysteines, four of which are involved in cross-linkages (Cys^{106}-Cys^{119} and Cys^{66}-Cys^{160}) and a free thiol Cys^{121}. β-Lg has been reported to release several bioactive peptides following hydrolysis by various enzymes,[16] and is therefore a protein of interest for hydrolysis experiments.

Different strategies may be followed in the mass spectrometry of protein hydrolysates:

- Hydrolysis may be followed by various chromatographic protocols aimed at sample cleaning and purification, which includes desalting on macroporous adsorption resins based on hydrophobicity of the peptide mixtures,[2] separation on the basis of size[22,23] or based on ionic properties[24,25] (see Chapter 2 for strategies and methods). Ultrafiltration and nanofiltration membranes have also been employed for peptide fractionation.[26,27] The process is repeated according to the strategy until a single peptide is obtained. MS(/MS) is performed on the collected single peptide to determine its mass and the amino acid composition

- In the case where the complete hydrolysate picture is the target, hydrolysis may be followed by liquid chromatography (LC) of the hydrolysates followed by ESI-MS(/MS). In this approach, no prior separation is necessary and the volume of information obtained is unwieldy, particularly where the substrate is not a single pure protein

- The third strategy is an online method, more or less designed along the stopped-flow system (see also Chapter 3)

In our work, we usually wish to capture the total breakdown pattern of the enzyme on purified substrates like β-Lg and/or α-lactalbumin (α-La).[7]

5.2 Hydrolysis Materials and Equipment

- Example with bovine β-Lactoglobulin (β-Lg):
  - Bovine β-Lg (protein content of 96%) prepared from whey protein isolate (WPI), a product from Fonterra Co-operative Group Ltd, Auckland, New Zealand
  - Bovine trypsin (EC 3.4.21.4) bought from Sigma-Aldrich, MO, USA
  - ⓘ Note if any residual chymotrypsin activity reported
  - ⓘ Treatment with tosyl-L-phenylalanine chloromethyl ketone (TPCK) alkylates His^{57} in the catalytic triad (Ser^{195}, His^{57} and Asp^{102}) of chymotrypsin, rendering it inactive
  - ⓘ Treatment with diphenylcarbamyl chloride (DPCC) also reduces chymotrypsin side activity
- Hydrolysis apparatus for the simulated pH-stat method:
  - An autotitrator (*e.g.* pH-Stat, TitroLine alpha plus, Schott AG, Mainz, Germany)
  - TitriSoft 2.5 hydrolysis management software (Schott AG, Mainz, Germany), run on a PC interfaced with the pH-stat equipment
  - ⓘ Different titrator manufacturers have their own software
  - ⓘ If you have no access to an autotitrator, you may titrate manually[28]

5.2.1 Enzymatic Hydrolysis of β-Lactoglobulin

5.2.1.1 Preparation of β-Lactoglobulin

Readily available β-Lg may be used. Alternatively, preparations from rennet whey protein concentrate (WPC) in which β-Lg is isolated by selective precipitation of α-La process according to Gésan-Guiziou *et al.*[29] and optimized by Tolkach and Kulozik[30] is an alternative.

5.2.1.2 The pH-Stat Controlled Hydrolysis

5.2.1.2.1 Sample Preparation

- Ensure you know the protein concentration in the substrate since the enzyme-to-substrate ratio (E/S) is calculated based on that
  - ⓘ For example, if your 10 g protein sample has 93.84% protein, the enzyme required for a 1% E/S (either weight/weight or volume/weight relationship) would be some 93.84 mg
- ➤ Weigh out 5–10 g β-Lg substrate into a 150 mL beaker
  - ⓘ You may dissolve it in the same or another beaker before introduction to the reactor (this is better if you want to work with accurate concentrations) or dissolve directly in the reactor. Either way, remember to adjust the pH to improve solubility
- With a spatula or spoon, introduce the powder into a beaker with the water for dissolving the substrate
- If you target about 100 mL reaction volume, make the solution in 70 mL water first. Use the remaining water to wash out the solution and make up the volume after transferring the dissolved substrate to a measuring cylinder

ⓘ Beware of cylinder volume error. Depending on how the substrate is dissolved, the final concentrations might change: for example, when dissolving 10 g in 100 mL water (total reaction volume would be >100 mL) or dissolving 10 g in 100 mL reaction volume (total reaction volume would be exact)

5.2.1.2.2 The pH-Stat Method.

The method is so called because the *pH* is kept *stat*ic or unchanged (actually corrected it is as it changes during hydrolysis).

- Connect the jacketed (glass or stainless steel) reactor to a thermostatted water bath
- Ensure the connecting tubes do not leak too much heat in order to save energy
    - ⓘ Where you do not have a jacketed reactor, a beaker may be immersed in a bigger one provided the bigger beaker has water reaching the level that would cover the hydrolysis reaction mixture
    - ⓘ Heating magnetic stirrers may also be used to supply the energy, ensure the level of the substrate is reached by the heating
- ➤ Dissolve in double distilled water or buffer to make reaction volume of 100 mL in the buffer, NaOH-buffered double-distilled water or suitable hydrolysis medium
    - ⓘ Protein solubility is influenced by pH, temperature and presence of denatured proteins which appear as flakes. If the substrate has low pH (*e.g.* acid whey proteins at pH 4.6) it forms lumps when introduced into water. To improve solubility, adjust the pH to above neutral and continue stirring
- ➤ You may use a 150 mL thermostatically controlled, well-stirred Schott Duran (or any other make available locally) jacketed-beaker glass batch reactor (We use reactors from HLL Landgraf Laborsysteme, Langenhagen, Germany.)
- ➤ Adjust the solution pH to the chosen value depending on the experimental design, *e.g.* pH 7.8
- Ensure the temperature is stable at the set value according to the experimental design, *e.g.* 37 °C. Usually there ought to be a difference between the water bath temperature and the reactor temperature
    - ⓘ Be sure to read the reactor temperature using a clean thermometer
    - ⓘ Do not rely on the titrator reading of the thermometer if you use a pH electrode that reads the temperature too
- ➤ If working with lyophilized enzymes, weigh out the enzyme for example into 1.5 mL cuvettes (placed into any long and narrow beaker or holder) containing about 0.5 mL water to trap the powder and prevent powder spreading in the neighbouring environment
- Pay attention to enzyme safety requirements which are provided with shipping materials or accessible on the internet; wear goggles and gloves, and avoid inhalation of the enzyme powder

ⓘ Keep the working area clean and beware of contamination by other chemicals and/or enzymes. Wipe the weighing scale clean in order to take care of work colleagues who might not be aware of the potentially harmful contamination

ⓘ The enzyme may be vortex mixed briefly if it does not dissolve readily

- Avoid holding dissolved enzyme in solution for longer times at pH and temperature conditions in which it is active because of the likelihood of autolysis or self-digestion

  ⓘ We usually use it immediately, within 5 min

  ⓘ If you must hold it for some time in solution, keep it refrigerated

➤ Ensure the solutions for pH adjustment are ready

- If you work at alkaline pH, manually add or autotitrate appropriate concentrations of NaOH (from 0.5 M to 2 M) to raise the decreasing pH

- When hydrolysis is performed at acid pH (*e.g.* for pepsin at pH 1.5–2.0), add 1–2 M HCl to regulate the pH because it rises during hydrolysis

  ⓘ Avoid higher concentrations of NaOH, in case the protein is denatured by alkali

5.2.1.2.3 Prior to Enzyme Addition

- Draw out a blank aliquot (for LC-MS/MS, 500 μL is enough)
- Mix that with the same acid used to stop the enzyme in subsequent Eppendorfs (in our case <80 μL of 1 M HCl)

  ⓘ Ensure the Eppendorfs are appropriately marked, *e.g.* for time t = 0 or blank, mark '0'

  ⓘ Subsequent Eppendorfs should also be marked. We work with 5 s (a sample drawn some 5 s after enzyme addition), 10 s, 15 s, 20 s, 30 s, 45 s, 1 min, 2.5 min, 5 min, 7.5 min and 10 min

  ⓘ If hydrolysis is performed for longer periods, mark the hydrolysis time appropriately

  ⓘ Arrange the Eppendorfs in a rack (Styroform used for packing delicate supplies can be used if a rack is unavailable; by making appropriate cut-outs the Eppendorfs are secured against falling over or confusion)

- It is important to add the enzyme-stopping acid/alkali into each Eppendorf before the commencement of hydrolysis

➤ If you use a stop-watch, reset it before the commencement of hydrolysis. If the hydrolysis process is connected to a PC and managed using software like Titrisoft, ensure the X-axis is configured to display time (in s or min) during hydrolysis

  ⓘ Check the pH and temperature to ensure they are stable at the experimental values, adjust if necessary and wait for stability. If the pH is lower than the target, add a few drops of 0.5 M or 1 M NaOH. If higher than the target, add a few drops of 0.5 M or 1 M HCl

  ⓘ Ensure the stirring is working well and the water-bath thermostat is functioning. Depending on the reactor size, a magnetic stirrer may be used

  ⓘ Listen for any noises from the thermostatted water-bath recirculation pump and fix any problems before you start the process

> If you wish to simulate the stop-flow method for monitoring peptide evolution over time by drawing aliquots of the reaction mixture then you need to prepare beforehand

- Enzyme reactions may be stopped instantly by changing the pH, heating the reactants or adding inhibitors to the enzyme
  - ⓘ Sometimes it is important to use pH-adjustment and heating if dealing with 'resilient' enzymes which are difficult to stop
- Remember you only stop the reaction, not necessarily denature the enzyme. We have noticed that even when the pH was suppressed to below pH 3.5, trypsin recovered some activity upon readjustment to about pH 8
  - ⓘ To stop reactions in the alkaline pH region (*e.g.* with trypsin) immediately, it is easier to add acid (usually 1–2 M HCl) to lower the pH to regions in which the enzyme has no residual activity
  - ⓘ To stop reactions in the acid pH region (*e.g.* with pepsin, EC 3.4.4.1, or acid protease A) add NaOH to raise to about pH 8.5
  - ⓘ Prepare Eppendorfs with appropriate volumes of 1 M or 2 M HCl beforehand
  - ⓘ Trypsin and chymotrypsin activity may be stopped instantly by introducing an inhibitor like Bowman–Birk inhibitor. However, a method must be considered to eliminate this in subsequent stages if applications are considered
  - ⓘ Enzyme activity may not be stopped by a single method like pH adjustment or temperature. A combination of both might be reasonable
- Beware of effects of harsh temperature and pH on the proteins and peptides owing to the likelihood of foulant formation and/or complexes with other chemicals in the reaction mixture

5.2.1.2.4 Start Enzymatic Reaction

- Introduce the enzyme solution (it is better to draw into a pipette than to pour directly into the reactor)
  - ⓘ If you use a pipette, rinse it several times by drawing the reactants and returning it to the reactor
- Watch out for the time to draw the first reaction mixture
  - ⓘ If you use manual titration of the pH-controlling solution from a burette, extra care is necessary not to panic during pH adjustment in order to avoid overshooting the limits of pH stability of the enzyme
- You might need to grease the burette tap to ensure it turns freely
- Ensure no gas bubbles are trapped at the burette outlet if you need accurate volume readings
> We work with trypsin and our work is normally controlled with addition of 0.5–2 M NaOH using an autotitrator for the pH-stat hydrolysis (pH-Stat, TitroLine alpha plus, Schott AG, Mainz, Germany)
  - ⓘ The hydrolysis may be managed using TitriSoft 2.5 (Schott AG, Mainz, Germany), run on a PC interfaced with the pH-stat equipment
  - ⓘ Ensure you set the parameters in Titrisoft Methods Centre. Set the pH, time of hydrolysis in minutes, choose the pH-stat method, pipette,

dosing solution, what to display on the axes during hydrolysis, *e.g.* time *versus* pH, titration curve (if hydrolysis is performed in alkaline pH, choose the increasing style)

ⓘ Switch to Titration Centre before you commence hydrolysis, especially if you use the titrator pH meter to monitor the pH before hydrolysis. See the system we use in our work in Figure 5.4 and in Cheison *et al.*[7]

5.2.1.2.5 DH Calculation. The DH is a useful index used to monitor protein digestion. The easiest way to calculate the DH is the pH-stat method.

➤ The amount of NaOH used to maintain the pH may be converted to instantaneous or final DH according to the pH-stat method using equation 5.2[1,10]

ⓘ Alternatively, the DH may also be calculated as nitrogen soluble in 5% trichloroacetic acid (TCA)[31] or the concentration of liberated NH_2 groups reacted with either ninhydrin[9] or TNBS[10,11]

➤ At the end of the reaction, keep the remaining reaction mixture with pH adjustment

ⓘ If it is not possible to analyse immediately, store the samples prior to analysis at $-18\,^{\circ}C$

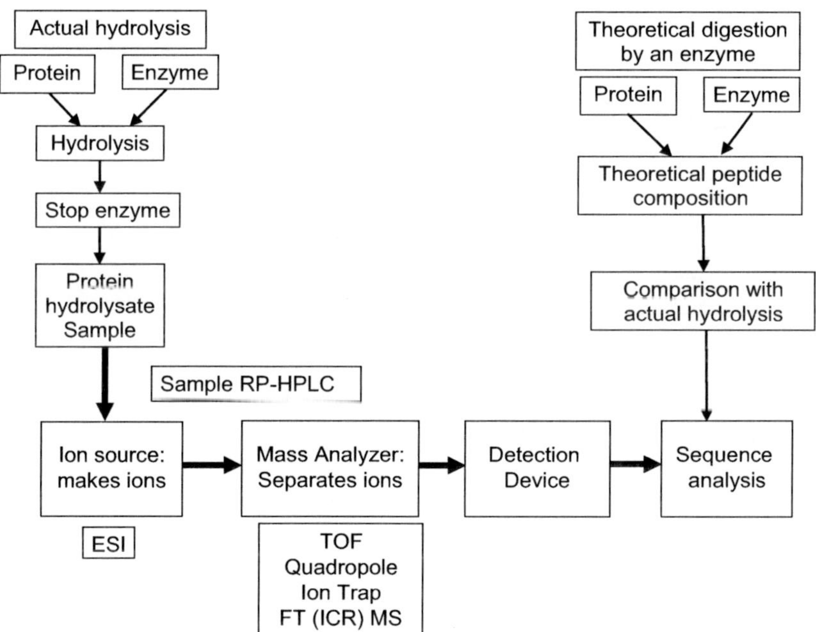

Figure 5.4 Typical workflow for the production and sample preparation for RP-HPLC hyphenated with ESI-MS(/MS) for hydrolysate separation and mass analysis. The hydrolysate mass spectra are compared to theoretical digests performed using the software if the enzyme specificity is known.

- Naturally, clean the equipment; take care to wear protective gloves when touching surfaces contaminated with the enzyme

5.2.2 Tips with Enzymes

Most of these tips can be found in the manufacturer's literature accompanying the purchased enzyme. Please read them carefully, it is easier to handle a crisis if you have read the precautions than to run around looking for literature in the middle of a problem.

With trypsin, however, remember:

- ⓘ It is an expensive enzyme. Use amounts that are economical. Where possible, re-use the enzyme with immobilization in membrane reactors,[32–34] immobilization in gels[35–37] or columns (see Chapter 6)
- ⓘ Storage of trypsin for a long period may lead to the development of side activity due to formation of pseudo-trypsin,[38] which is blamed for the cleavage of chymotrypsin-specific bonds like Tyr-Ser.[7,39] Buy only those amounts you can deplete if supply is steady
- Pay attention to storage conditions:
  - ⓘ Usually keep the container tightly closed and frozen at $-18\,^\circ$C or lower, depending on enzyme stability at the chosen temperature
  - ⓘ Mind your safety and those of your workmates; do not leave enzyme spills uncleaned
- Trypsin lyophilizate may be carried by strong air currents
  - ⓘ Ensure the weighing room is free from strong air blowers and air currents
  - ⓘ Wetting a paper tissue with clean tap water and using it to wipe off the powder is sufficient
- It is important to read safety precautions and/or Material Safety Data Sheets (MSDS) documentation in order to get accurate information on how to handle cases of accidental inhalation
  - ⓘ If the enzyme comes in contact with eyes *etc.*, wash with excess tap water

5.3 LC-ESI-TOF/MS Spectra Equipment and Methods

5.3.1 Equipment

- An Agilent rapid resolution high performance liquid chromatography (HPLC) system (series 1200, Waldbronn, Germany) in combination with a time-of-flight mass spectrometer (G1969A; 6210 TOF LC/MS, Santa Clara, CA, USA)
  - ➤ In our work, the apparatus consists of:
    - A binary capillary pump (G1312B)
    - An isocratic pump (G1310A)
    - A degasser unit (G1379B)

- ■ An autosampler (G1367C)
- ■ A column in a thermostat set to 50 °C (G1316B)
- The mass-selective detector was connected by an electrospray ionization (ESI) interface (G3251A)
- The HPLC, ion source and mass spectrometric detector were controlled and data was analysed by MassHunter Workstation software (A.02.02, Agilent, Waldbronn, Germany)
- Optimize the tube connections and connections for low dead volume performance

Liquid chromatography LC-ESI-TOF-MS may be used as reported already.[7] For example:

- ➢ Prontosil 120-3-C_{18} column (dimensions: 100 mm × 2 mm, particle size 3 μm, Bischoff Chromatography, Leonberg, Germany)
- ➢ Sample injection volume of 20 μL
- ➢ Mobile phase was $MeOH/H_2O$ at a flow-rate of 200 μL/min

5.3.2 Peptide Mass Fingerprinting

This approach is a bottom up rather than top down analysis, since the amino acid sequence of the substrate is known already.

The peptide masses can be checked with:

- ➢ FindPept database (http://ca.expasy.org/tools/findpept.html) for peptide searches including specific and non-specific cleavage
- ➢ PROWL's Peptidemap (http://prowl.rockefeller.edu/prowl/peptidemap.html) which is good for peptide search and disulfide-linked peptide identification
  - ⓘ If you use Bruker Daltoniks equipment and/or BioTools software, the online search for peptide fingerprints may not be necessary
  - ⓘ A simple workflow is provided by Bruker in which a theoretical digest of the substrate by an enzyme like trypsin in the Sequence Editor may be combined with the acquired spectra in flexAnalysis and sent in the same way to BioTools

5.4 Data Analysis

Depending on the purity of the substrate, it should be relatively easy to map out the peptide sources in the amino acid sequence of the substrate.

5.4.1 Tips with MS Data from Trypsin Hydrolysates

- Watch out for adducts of Na^+ (shifts of $+22$ for single-charged or $+11$ for double-charged states) and other modifications which are listed at http://ca.expasy.org/tools/findmod/findmod_masses.html
- Check out common adducts with the versatile Adduct Calculator available at http://fiehnlab.ucdavis.edu/staff/kind/Metabolomics/MS-Adduct-Calculator

Table 5.1 Peptides undetected but probably released by trypsin hydrolysis of
β-Lg

Fraction	Peptide[a]	AM (Da)[b]	Calculated pI[c]	H_oAA[d] (%)
f41–70[e] (29-mer)	VYVEELKPTPEGDLEILLQK EGGECAQKK	3244.7	4.24	39.9
f76–91 (16-mer)	TKIPAVWKIDALNENK	1840.1	9.45	49.8
f102–124 (23-mer)	YLLFCMENSAEPEQSLA CQCLVR	2648.1	3.98	43.3
f149–162 (14-mer)	LSFNPTQLEEQCHI	1658.8	4.25	35.6

[a]Single-letter amino acid symbols used
[b]Average mass.
[c]Calculated using Peptide Property Calculator available at https://www.genscript.com/ssl-bin/site2/peptide_calculation.cgi (accessed 8 Nov 2010)
[d]Content of hydrophobic amino acids, H_oAA (alanine (A), isoleucine (I), leucine (L), phenylalanine (F), Proline (P), tryptophan (W), tyrosine (Y) and valine (V)) where present.
[e]Number of amino acids given as '-mer'

- Watch out for some common peptide bonds like Tyr-Ser[7,40–42] which are commonly cleaved due to non-specific trypsin activity attributed to pseudo-trypsin activity[43,44] or presence of residual chymotrypsin in the enzyme preparation, although this was detected even with TPCK-treated trypsin.[39]
- See Chapter 10 for useful hints with LC-MS
- Look out for missing peptides in ESI-MS especially those composed of high amounts of hydrophobic amino acids (Table 5.1)[7]

5.5   Conclusions

Protein hydrolysis is a process that is sensitive to the hydrolysis environmental conditions such as pH and temperature as well as the ionic strength of the reaction buffer. These conditions may impact on enzyme specificity in more ways than its effect on the 'speed' of reaction.

It is important while working with enzymes to observe one's own safety and that of others. It is important to handle enzymes with the conscious knowledge of their potential harm to health.

Mass spectrometry of hydrolysates is easier when working with purified proteins than with mixed proteins. In addition, the hydrolysis buffer may interfere with the mass spectrometry of peptides due to the formation of peptide-salt ion adducts.

References

1. J. Adler-Nissen, *Enzymic Hydrolysis of Food Proteins*, Elsevier Applied Science Publishers, London, 1986.
2. S. C. Cheison, Z. Wang and S.-Y. Xu, *Int. J. Food Sci. Technol.*, 2007, **42**, 1228.

3. J. J. Pahud, J. C. Monti and R. Jost, *J. Pediatr. Gastroenterol. Nutr.*, 1985, **4**, 408.
4. H. G. Kristinsson and B. A. Rasco, *Process Biochem.*, 2000, **36**, 131.
5. M. I. Mahmoud, W. T. Malone and C. T. Cordle, *J. Food Sci.*, 1992, **57**, 1223.
6. C. van der Ven, H. Gruppen, D. B. A. de Bont and A. G. J. Voragen, *J. Agric. Food Chem.*, 2002, **50**, 2938.
7. S. C. Cheison, M. Schmitt, E. Leeb, T. Letzel and U. Kulozik, *Food Chem.*, 2010, **121**, 457.
8. F. Camacho, P. González-Tello, M. P. Páez-Dueñas, E. M. Guadix and A. Guadix, *J. Dairy Res.*, 2001, **68**, 251.
9. R. McGrath, *Anal. Biochem.*, 1972, **49**, 95.
10. S. C. Cheison, S.-B. Zhang, Z. Wang and S.-Y. Xu, *Food Res. Int.*, 2009, **42**, 91.
11. J. Adler-Nissen, *J. Agric. Food Chem.*, 1979, **27**, 1256.
12. D. Spellman, E. McEvoy, G. O'Cuinn and R. J. FitzGerald, *Int. Dairy J.*, 2003, **13**, 447.
13. J. M. Ezquerra, F. L. Garcia-Carreno, R. Civera and N. F. Haard, *Aquaculture*, 1997, **157**, 251.
14. J. V. Olsen, S.-E. Ong and M. Mann, *Mol. Cell Proteomics*, 2004, **3**, 608.
15. M. Careri and A. Mangia, *J. Chromatogr. A*, 2003, **1000**, 609.
16. B. Hernández-Ledesma, I. Recio and L. Amigo, *Amino Acids*, 2008, **35**, 257.
17. F. Zhong, X. Zhang, J. Ma and C. F. Shoemaker, *Food Res. Int.*, 2007, **40**, 756.
18. J. B. Fenn, M. Mann, C. K. Meng, S. F. Wong and C. M. Whitehouse, *Science*, 1989, **246**, 64.
19. F. Hillenkamp and M. Karas, *Methods Enzymol.*, 1990, **193**, 280.
20. M. Karas and F. Hillenkamp, *Anal. Chem.*, 1988, **60**, 2299.
21. J. P. Hill, M. J. Boland, L. K. Creamer, S. G. Anema, D. E. Otter, G. R. Paterson, R. Lowe, R. L. Motion and W. C. Thresher, *ACS Symposium Series 650*, p. 281. American Chemical Society, Washington, DC, 1996.
22. D. Spellman, P. Kenny, G. O'Cuinn and R. J. FitzGerald, *J. Agric. Food Chem.*, 2005, **53**, 1258.
23. P. W. Caessens, S. Visser, H. Gruppen and A. G. Voragen, *J. Agric. Food Chem.*, 1999, **47**, 2973.
24. R. M. Barros and F. X. Malcata, *J. Agric. Food Chem.*, 2002, **50**, 4347.
25. M. R. Guo, P. F. Fox, A. Flynn and P. S. Kindstedt, *J. Dairy Sci.*, 1995, **78**, 2336.
26. A. Pihlanto-Leppälä, P. Koskinen, K. Piilola, T. Tupasela and H. Korhonen, *J. Dairy Res.*, 2000, **67**, 53.
27. A. Pihlanto-Leppälä, P. Marnila, L. Hubert, T. Rokka, H. J. T. Korhonen and M. Karp, *J. Appl. Microbiol.*, 1999, **87**, 540.
28. S. C. Cheison, Z. Wang and S.-Y. Xu, *Int. Dairy J.*, 2007, **17**, 393.
29. G. Gésan-Guiziou, G. Daufin, M. Timmer, D. Allersma and C. Van Der Horst, *J. Dairy Res.*, 1999, **66**, 225.

30. A. Tolkach and U. Kulozik, *Lait*, 2007, **87**, 301.
31. S. C. Cheison, Z. Wang and S.-Y. Xu, *J. Food Eng.*, 2007, **80**, 1134.
32. S. C. Cheison, Z. Wang and S.-Y. Xu, *J. Membr. Sci.*, 2006, **283**, 45.
33. A. Guadix, F. Camacho and E. M. Guadix, *J. Food Eng.*, 2006, **72**, 398.
34. S. C. Cheison, Z. Wang and S. Y. Xu, *J. Agric. Food Chem.*, 2007, **55**, 3896.
35. J. Gobom, E. Nordhoff, R. Ekman and P. Roepstorff, *Int. J. Mass Spectrom. Ion Processes*, 1997, **169–170**, 153.
36. X. L. Huang, G. L. Catignani and H. E. Swaisgood, *J. Biotechnol.*, 1997, **53**, 21.
37. R. M. Blanco, J. J. Calvete and J. Guisán, *Enzyme Microb. Technol.*, 1989, **11**, 353.
38. B. Keil, *Specificity of Proteolysis*, Springer-Verlag, New York, 1992.
39. I. Sélo, G. Clément, H. Bernard, J.-M. Chatel, C. Créminon, G. Peltre and J.-M. Wal, *Clin. Exp. Allergy*, 1999, **29**, 1055.
40. F. Maynard, A. Weingand, J. Hau and R. Jost, *Int. Dairy J.*, 1998, **8**, 125.
41. L. J. Greene and J. S. Giordano, Jr., *J. Biol. Chem.*, 1969, **244**, 285.
42. T. Asao, I. Tsuji, M. Tashiro, K. Iwami and F. Ibuki, *Biosci. Biotechnol. Biochem.*, 1992, **56**, 521.
43. V. Keil-Dlouhá, N. Zylber, J. M. Imhoff, N. T. Tong and B. Keil, *FEBS Lett.*, 1971, **16**, 291.
44. V. Keil-Dlouhá, N. Zylber, N. T. Tong and B. Keil, *FEBS Lett.*, 1971, **16**, 287.

On-line Protein Digestion in Combination with Chromatographic Separation and Mass Spectrometric Detection

S. JOHANNES HOOS AND WILFRIED M.A. NIESSEN

VU University, Faculty of Sciences, BioMolecular Analysis Group, De Boelelaan 1083, 1081 HV, Amsterdam, The Netherlands

6.1 Introduction

The combination of protein digestion, separation prior to or after digestion, and detection using mass spectrometry (MS) is of great importance in many research areas. It is frequently applied using a variety of approaches, *e.g.* proteomics research, studies on protein primary structure, target protein analysis, protein biomarker studies and analysis of protein modifications. Peptides resulting from a proteolytic digestion can provide important information on the proteins, which may be difficult or impossible to obtain otherwise from the intact protein molecule.

Although the detection of peptides can be performed using a wide variety of techniques like UV detection, (laser-induced) fluorescence detection or detection *via* radioactive labelling, MS has become a major detection principle for in-depth analysis of proteins and peptides. The reason for this may be found in the fact that mass analysers are capable of providing specific information on

RSC Chromatography Monographs No. 15
Protein and Peptide Analysis by LC-MS: Experimental Strategies
Edited by Thomas Letzel

Published by the Royal Society of Chemistry, www.rsc.org

the molecules of interest, assuming proper and reproducible ionization behaviour of the target molecule(s). Since MS is based on the detection of analyte ions, the target proteins have to be ionized before introduction into the mass analyser. Three main methods are available for the ionization of biomolecules: fast atom bombardment (FAB), electrospray ionization (ESI), and matrix-assisted laser desorption ionization (MALDI).[1] Although any of these ionization methods may be used for peptide and protein ionization, ESI-MS has become popular for on-line analysis with separation methods for several reasons. First, this method supports the continuous introduction of sample by a carrier flow or by a column effluent. This enables direct analysis of proteins or peptides eluting from a chromatographic separation column. Second, ESI-MS (like MALDI-MS) allows the analysis of very low quantities of the analyte. Furthermore, because MS determines the mass-to-charge ratio (m/z) of the analyte ions and ESI-MS provides the ability to add multiple charges to a molecule, it enhances the ability to detect larger biomolecules with instruments featuring limited m/z range. In addition, it provides the possibility to perform MS^n experiments for structure elucidation, amino acid sequencing and biomolecule identification. However, these positive aspects of ESI-MS are only applicable in a molecular mass range up to 10 000 Da whereas a typical protein features a molecular mass of several thousand up to hundreds of thousands (10^4–10^5 Da). The reason for this is the limited ability of the ionization method to efficiently ionize such large molecules and the limited resolution of the mass analyser when used with higher m/z values. Therefore, proteins are commonly identified and analysed by ESI-MS as smaller fragments (peptides), that are enzymatically or chemically derived from the original protein. The use of these protein fragments circumvents the formation of a multiple charge distribution over a wide m/z range observed in ESI-MS protein analysis and enables analysis in an appropriate m/z range. Several types of mass analysers have been developed for m/z determination and/or subsequent characterization by MS^n, including (linear) ion trap, quadrupole–time-of-flight (Q-TOF), quadrupole–linear-ion-trap and linear-ion-trap orbitrap instruments. Further instrumental details can be found in Chapter 2. This chapter reviews various methods to achieve protein proteolysis on-line with separation methods and mass spectrometry.

6.2   Proteolysis of Proteins

Peptides can be obtained by exposure of the protein of interest to proteolysis. Proteolysis is the directed degradation of proteins by enzymes (enzymatic digestion) or by non-enzymatic reactions (chemical digestion). Proteolysis by enzymatic cleavage is most frequently used since this type of digestion can be performed under relatively mild conditions. Numerous proteolytic enzymes are available from different organisms and every species exhibits a specific enzymatic activity. Chemical digestion is generally less specific and does not yield relatively predictable cleavage:[2] only a small number of chemical digestion methods find suitable application within a proteomics approach.[3]

A commonly used strategy to facilitate efficient digestion of proteins is denaturation and reduction. The denaturation process is applied to disrupt the native tertiary protein structure, especially by cleaving S-S bridges between cysteine moieties. This can be achieved by heat, extreme pH values or by addition of chaotropic agents such as urea, guanidinium chloride or sodium perchlorate. Frequently, alkylation (*e.g.* with iodoacetamide) is performed after reduction (*e.g.* with dithiothreitol, DTT) to prevent formation of new S-S bonds. In this way, the protein is prevented from refolding. After removal of the denaturation and reduction agents, a very broad range of proteases can be applied employing several strategies. After digestion has been applied under suitable conditions, the resulting peptide mix can be analysed.

As an example, a general protocol for off-line trypsin digestion of a protein typically looks like the following:

- Resuspend the protein in 6 M urea and 50 mM Tris buffer at pH 7.8 to a concentration of 10 mg/mL
- Transfer 100 μL of the solution to a new 1.5 mL tube
- Add 5 μL of reducing agent (200 mM DTT in 50 mM Tris solution, pH 7.8). Vortex and incubate for 30 minutes at room temperature (typically, 20–25 °C)
- Add 20 μL of alkylation reagent (100 mM iodoacetamide in 50 mM ammonium bicarbonate solution, pH 8.0. Note: the iodoacetamide solution must be stored in the dark). Vortex and incubate for 60 minutes at room temperature protected from light
- Quench the unreacted iodoacetamide by adding 15 μL of reducing agent prepared as in step 3
- Reduce the urea concentration by diluting the reaction mixture with 775 μL of water. Alternatively, the sample may be desalted, *e.g.* by using a size-exclusion gravity column. (In this case, use 50 mM Tris solution at pH 7.8 as the mobile phase.)
- Add 50 μL of trypsin solution (1 mg/mL with a specific activity higher than 500 U/mg in 50 mM ammonium bicarbonate and 1 mM $CaCl_2$ solution, pH 8.0). Vortex the reaction mixture and incubate overnight at 37 °C
- After, *e.g.*, 18 hours, stop the reaction by adding formic acid to 1% (v/v, final concentration)
- If urea was removed in step 6, inject the mixture into a high-performance liquid chromatography (HPLC) system and analyse the sample. Otherwise, clean up the peptide content of the reaction mixture by performing a solid-phase extraction (SPE) clean-up, *e.g.* using a C18 spin column, evaporate the eluate to dryness, resuspend the sample in water or mobile phase to a suitable concentration, *e.g.*, to 0.1 mg/mL of the initial protein concentration, inject onto an HPLC system and analyse the sample
- When the sample is to be analyzed later, evaporate to dryness and store at −20 °C. It is also possible to store the digested solution mix at −80 °C until analysis

A large variation of proteolytic enzymes is available and can be used for site-specific digestion of proteins into smaller peptide fragments. The choice of enzyme strongly depends on the aim of the investigation. Trypsin is mostly used for proteolysis because of its well-defined properties such as specificity, stability and activity. The use of trypsin enables high turnover rates and predictable peptides due to a low degree of miscleavages and relatively specific cleavage at the carboxyl side of the amino acids Arg and Lys (except when linked to a Pro). These amino acids are present in an average rate of one residue per every 10–12 amino acids, and the resulting tryptic peptides typically have a mass between 800 and 2000 Da. Other frequently used enzymes are pepsin, chymotrypsin, endoproteinases such as Lys-C, Asp-N, Glu-C,[4] pronase[5] and protease K. Chymotrypsin cleaves under similar conditions as trypsin at the C-terminal site of Phe, Tyr, Trp, Leu and Met (except prior to a Pro). Furthermore, pepsin can be applied as a general protease for characterization of proteins; it cleaves relatively specifically at the C-terminus of Phe, Met, Leu and Trp. Pronase has proteolytic activity to both denatured and native proteins, and comprises various types of proteases. Therefore, pronase is relatively non-specific and can (after sufficient reaction time) break down proteins to their individual amino acids.

The proteolytic enzymes can be used in different ways. Most frequently, the protein digestion is performed in batch (*i.e.* in an off-line procedure) prior to further analysis by chromatography and/or (tandem) MS and possibly other technologies. The off-line procedure is a relatively laborious method that allows for the manual or semi-automated sample pretreatment and almost unlimited digestion time, which in turn can be time-consuming. After subjecting the sample to the desired pretreatment protocol (purification and/or denaturation and reduction), the enzyme of choice can be added to the sample at chosen conditions. Digestion is often performed overnight. If the digestion has been successful, liquid chromatography and ESI-MS analysis, or alternatively direct MALDI-MS, provides the desired peptide map of the proteins. Off-line sample pretreatment and digestion of protein samples has some additional drawbacks. Since off-line sample treatment involves manual sample handling such as pipetting or centrifugation, sample loss is frequently observed because the initial sample volume cannot be fully recovered, which in turn negatively affects detection limits. Also, there is an increased risk of contamination of the sample. In addition, off-line digestion procedures can be time consuming or require the use of sophisticated equipment for the automated sample handling steps. This often prevents medium- or high-throughput analysis of samples.

Alternatively, there exist a number of on-line protein digestion methods, such as immobilized enzyme reactions (IMERs) and continuous-flow reactors. These approaches can be applied either in precolumn or postcolumn mode. These are further discussed in section 6.4.

6.3 Immobilized Enzyme Reactors

The introduction of IMERs has been of great importance for the application of proteolysis in automated and high-throughput analysis. The first IMER was a reactor where trypsin was immobilized onto the inner surface of a capillary.[6] Later, IMERs based on immobilization of proteases on beads, membranes, microchannels, microchips, monoliths and silica particles were introduced,[7] which enabled a very wide range of applications for on-line protein digestion.

If the required protease is already available in an IMER format, its utilization is relatively simple. The protein or protein mixture is exposed to the immobilized enzyme by transporting the sample to the active material. After this, digestion can occur for a chosen period of time. However, if the required protease is not commercially available in an immobilized format, one has to develop and apply an immobilization procedure and subsequently characterize the resulting IMER, including long-term stability testing.

Whereas off-line digestion is performed for up to 24 hours, on-line digestion is relatively fast, *i.e.* several seconds up to 2 hours,[8] depending on the reactor activity and the selected conditions. Of all types of IMERs, the column format is mostly used because it can be operated completely without manual sample handling. Several media have been developed varying for example in non-specific absorption behaviour, long-term stability, thermal stability and enzyme density.

Whereas free enzymes can be characterized by the amount of enzyme activity that catalyses the transformation of one mole of substrate per second at 25 °C under optimal experimental conditions, the characterization of immobilized enzyme reactors requires additional parameters. The International Union of Pure and Applied Chemistry (IUPAC) requires characterization of IMERs by the percentage of immobilized enzyme, the enzyme activity after immobilization, the time and temperature stability of the immobilized enzyme, the pH optimum and the K_m value[9] (*i.e.* the substrate concentration at which the turnover rate of the reaction reaches half of its maximum) for appropriate substrates. To characterize an IMER for digestion of a given substrate, the scientist must keep additional issues in mind for the specific application. For example, the enzyme carrier must allow the protein to be easily transported to the active sites of the support, *e.g.* when porous material is used. Furthermore, non-specific interactions of the support with matrix compounds, protein or resulting peptides can seriously hinder the enzymatic reaction. Finally, the material of the column housing as well as the column material itself must be stable under the prevailing conditions, such as high temperature and/or elevated system back-pressure, due for instance to the use of directly connected chromatography columns.

Several types of immobilized enzymes are commercially available in various column dimensions. IMERs based on trypsin are most commonly used because they are well characterized and can be operated under mild conditions. As an alternative, IMERs based on chymotrypsin can be used. Immobilized pepsin is used frequently for the generation of F(ab')2 fragments from antibodies.[5] Other IMERs frequently used are based on endoproteinases (such as Lys-C, Asp-N, Glu-C) and pronase.

Immobilization has been performed on several kinds of materials for the use in a batch or in a column format. Immobilization on to beads like magnetic latex particles, paramagnetic particles, chitin, chitosan[10] or polystyrene beads[11] enables the exposure of enzymes to any kind of crude sample without the concern of clogging an IMER column. After digestion, the enzyme can be removed by spinning down the beads or (in the case of enzymes immobilized to paramagnetic beads) by applying a magnetic field. In an ideal case, the digested sample will be completely free of enzymes, and the enzyme-loaded beads can be re-used for further experiments after washing.

The use of IMERs in a column format requires completely homogeneous samples without solid particles present since stirring or homogenization is not possible in this format and the column might become clogged. Frequently used solid support materials utilized in the column format are agarose, methacrylate resin, porous glass spheres, polymeric or silica-based particles. The immobilization itself can be established *via* a covalent bond by a reactive group or *via* physical adsorption on to the material. Immobilization *via* a covalent bond is mostly often used, and includes chemical reaction principles *via* amino, epoxy, carboxyl, thiol and phenolic groups. The mode of linkage to the protein to be immobilized determines the binding reaction, such as diazetonization, amide bond formation, arylation or Schiff's base formation.

In a column format, the digestion time can be varied in two different ways. After the protein to be digested has been transferred into the IMER, the carrier flow can be stopped for a chosen amount of time. Application of a stop-flow mode enables almost unlimited digestion time, which in turn decreases sample throughput. This approach is especially suitable for detailed protein identification purposes, where a high degree of digestion of the protein into peptides is required.[12]

Another approach to vary digestion time when using an IMER in a column format is the flow digestion mode. Here, the digestion time can be varied by altering the linear flow rate, which in turn influences the time that the protein is exposed to the immobilized enzymes on the column. This approach is very suitable for quantification purposes, where constant digestion time of the entire amount of protein is required without the need to synchronize the interruption of the column flow with the elution time or appearance time of the protein in the digestion column.

IMERs can be successfully used provided that the cleavage sites of the targeted protein or protein mixture are accessible for the immobilized protease. Steric hindrance results in low conversion rates, or can simply hinder the digestion procedure. Application of suitable sample pretreatment steps may be inevitable for analysis where high sequence coverage is required. Many approaches and products are available to increase the representation of the complete set of possible peptides of a protein after digestion. However, even in favourable cases, routine LC-MS analysis detects protein digest peptides covering only 50–90% of its sequence,[13] which is generally sufficient for protein characterization or identification.

6.4 Methods Employing IMERs in Hyphenated Analytical Systems

When an IMER is used under conditions that are ESI-MS compatible, it can be directly hyphenated to MS detection in an on-line strategy. Such a simple approach can yield precious information on the identity of the injected proteins with relatively high sample throughput. This approach can be used when the complete peptide map is not required, *e.g.* in the case of protein identification by means of peptide database searching. In this strategy, the IMER is attached directly to an ESI-MS instrument and the protein samples are introduced to the IMER for a given digestion reaction time. After this, the peptides are coeluted to the detector which then analyses the peptides originating from the injected protein by MS^n. Digestion conditions can be optimized by varying parameters like reaction temperature, reaction time and use of different organic modifiers and buffer ingredients. However, compatibility with ESI-MS needs to be kept in mind. Alternatively, the IMER effluent can be adapted by using a make-up flow which may improve compatibility with ESI-MS detection, *e.g.* by adding a moderate (10–30 v/v final concentration) amount of an organic modifier (acetonitrile or methanol) to the effluent to improve ionization efficiency.

The (quantitative) analysis of proteins present in different matrices by means of peptides generated in an IMER may complicate this approach. Therefore, a sample preparation step, *e.g.* a reversed-phase (RP) column, may be introduced prior to the IMER in a postseparation or postcolumn digestion approach. A six-port valve can be introduced to desalt the sample and elute ESI-MS incompatible components to waste before introduction of the protein to the IMER. This can be done by applying a gradient of low to high organic modifier content. However, most enzymes exhibit limited compatibility with high percentages of organic modifiers. A counter-gradient, introduced between the RP column and the IMER, may be used to reduce or counterbalance the organic modifier content.

Sample throughput of such a system simply depends on a combination of the digestion time, (if required) the separation time of the sample pretreatment method prior to digestion, and the re-equilibration of the system. When a higher sample throughput is required, multiple parallel columns may be used for sample preparation. An example is shown in Figure 6.1, where separation is performed and the effluent is digested and analysed by ESI-MS. While the first column equilibrates and desalts the sample injected *via* the isocratic pump, the second column can perform the separation of the intact proteins present in another sample with the gradient pump.

A drawback of this strategy is the limited capability to differentiate between similar proteins which are successfully differentiated by means of their peptide fragments. A full peptide map is required to visualize the complete protein composition. Complete visualization of the peptide map in turn requires that the peptides present in a peptide mixture are all completely ionized, which is not possible when using direct infusion without separation due to ionization suppression effects of the coeluting peptides or other sample constituents. In practice, some peptides from the peptide map may not be observed because of poor ionization properties, but this is a general issue, independent of the way the peptide map is generated. Furthermore, most enzymes require the presence of high salt concentrations, non-volatile buffer constituents or cofactors that are not compatible with the ESI-MS technique.

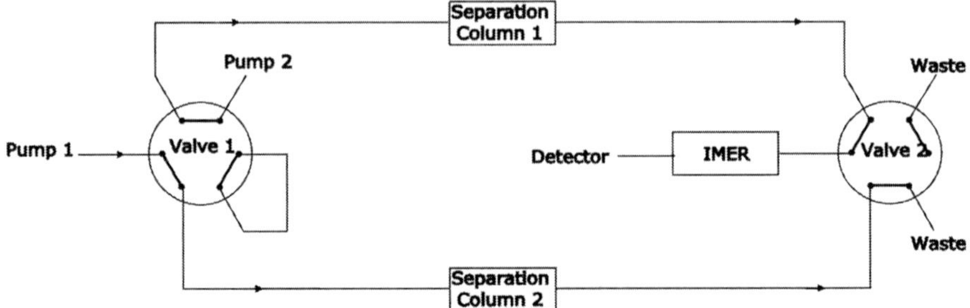

Figure 6.1 Setup of a postcolumn digestion approach employing two columns for increased sample throughput. Isocratic pump 1 transports an injected sample on to separation column 2, the intact protein (mix) is captured on the column and undesired matrix contents (*e.g.* salts) can be flushed to waste. After this, valve 1 is switched and the sample is eluted over the IMER to the detector by switching valve 2. Meanwhile, the next sample can be injected and desalted *via* column 1.

To circumvent ionization suppression effects of coeluting analytes and/or buffer constituents, the peptides can be separated downstream to the IMER in a preseparation or precolumn digestion strategy. When the conditions are carefully selected, the peptides can be reconcentrated, separated, efficiently ionized and subsequently detected. The preseparation digestion approach allows for linkage between non-ESI-MS compatible conditions used for the enzyme reaction and ESI-MS compatible separation and detection of peptides. Such a system (in this example an RP column and a trypsin IMER) may work as follows (see also Figure 6.2):

- An isocratic pump (Pump 1) is connected to an IMER *via* an injector. The outlet of the IMER is connected to a six-port valve (Valve 1) and the effluent of the IMER is diverted to waste. A solid-phase extraction (SPE) column is connected next to the open valve position where the IMER is connected. Also, a gradient pump (Pump 2) is connected to the same valve and this pump delivers the solvent for conditioning the SPE column. The effluent of the SPE column is either flushed to waste or pumped over an analytical column to the detector, which is controlled by valve 2. For the time being, the analytical column is conditioned by an isocratic pump
- After all the columns are equilibrated, a protein sample is injected on to the IMER and digested at given conditions, *e.g.* temperature, required buffer system, flow rate and pH. When the peptides start to elute from the IMER, these are captured on the SPE column by switching the first six-port valve (Valve 1). During this process, the SPE column effluent is diverted to waste *via* Valve 2
- After the peptides are completely eluted to the SPE column, the first six-port valve is switched back, enabling desalting of the captured peptides with gradient Pump 2 at low organic modifier concentration

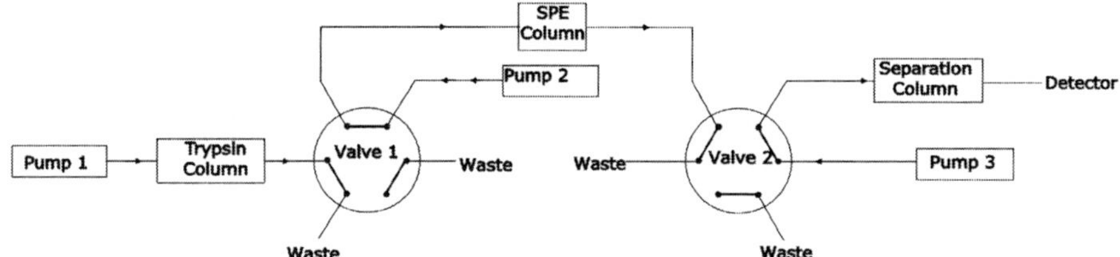

Figure 6.2 Setup of an on-line precolumn digestion system. Pump 1 is an isocratic pump where the sample is injected onto the trypsin column. The digested protein is transported to the SPE column and subsequently desalted by switching valve 1, connecting the SPE column to gradient pump 2. After the sample is desalted, gradient pump 2 runs a gradient over the desalted peptides on the SPE column, the separation column and the detector by switching valve 2. After this, valve 2 is switched back and pump 3 re-equilibrates the LC-MS. While pump 2 re-equilibrates the SPE column, the next sample can already be digested.

- Finally, Valve 2 is switched, connecting the gradient pump to the SPE column and the separation column. A gradient of low to high organic modifier content elutes desalted peptides from the SPE column to enable separation of the captured peptides, where the peptides can then be ionized and detected by ESI-MS. During this step, the IMER can be cleaned and re-equilibrated for the next injection
- After the gradient is completed, Valve 2 is switched back, the SPE column can be re-equilibrated by Pump 2, the next sample can be digested, captured and desalted, while the separation column is reconditioned using low organic modifier solution delivered by Pump 3

The sample throughput of such a method depends on the total time required to either digest or separate the peptides (whichever takes longer) and the time needed for the re-equilibration of the SPE cartridge, which can be minimized by using a higher flow rate during this process. Although the digested sample could be directly captured on the separation column without using the SPE cartridge, sample throughput is increased by introducing an SPE cartridge before the separation column, because the SPE cartridge in general requires less time for re-equilibration, and more readily enables desalting. Also, constant pressure can be maintained on the analytical column, which is important for column lifetime. Most important, however, is that the reduced back-pressure of short SPE cartridges is directly compatible with most IMER housings and IMER packing materials, which are not always resistant to high pressure.

Many approaches can be applied in on-line analytical separation protocols after a precolumn digestion strategy. Examples include size exclusion chromatography (SEC), reversed-phase liquid chromatography (RP-LC), ion exchange chromatography (IEC) or capillary electrophoresis (CE) and even on-line selection of specific peptides *via* immobilized metal affinity chromatography (IMAC),[14] immuno-[15] and lectin affinity[16,17] chromatography. However, the most widely applied approach is RP-LC,[18] which combines the ability to reconcentrate, efficiently separate and directly detect by ESI-MS.

If complex samples need to be analysed, *e.g.* biological matrices such as plasma or urine, sample pretreatment can be automated, which makes the resulting highly hyphenated system more complex, but further reduces manual sample handling. Several processes can be automated without manual interference, including the denaturation/alkylation processes prior to digestion. Sample clean-up before the introduction of the proteins to the IMER reduces matrix effects during the digestion procedure. As an example, it is possible to selectively clean up a protein using immunoaffinity chromatography (IAC), which is directly coupled to an IMER for digestion. The resulting peptides are re-concentrated by SPE, desalted and separated by RP-LC prior to ESI-MS^{n} detection.[19]

Before setting up such a system, it is highly recommended to carefully evaluate the individual steps of the process for linearity, reproducibility, carry-over effects and recovery. After this, one may start to set up the system from the detection end backwards, *i.e.* first performing the separation of an off-line digested and desalted sample in order to evaluate the analytical variables. Secondly, set up the on-line desalting and re-evaluate the analytical variables. Next, incorporate the on-line digestion, vary conditions calculating turn-over rates and optimizing recoveries. When any form of on-line sample pretreatment is required, this process can subsequently be attached in front of the on-line digestion approach.

6.5   Methods Employing In-Solution Digestion in Continuous-Flow Reactors

The use of IMERs is limited by the ability to immobilize enzymes on a suitable carrier without affecting their effectiveness. Furthermore, the application of IMERs becomes infeasible when a large variety of different enzymes has to be screened for their potential to digest a certain substrate. In the latter case, immobilization is not an option and less laborious methods might be favoured.

A possible strategy to evaluate the potential of enzymes for specific proteolytic reactions is the use of in-batch continuous-flow digestion analysis of a reaction mixture. In such an approach, the protein of interest is mixed with the enzyme and the reaction mixture is directly and continuously infused into a mass spectrometer. The composition of the reaction mixture can then be monitored continuously, provided that the conditions are ESI-MS compatible. In a similar approach, a reaction mixture can be analysed at given time points by performing flow-injection experiments using a suitable flow-carrier. In both approaches, it is very important to perform control experiments to evaluate the background of the carrier flow, the contribution of the enzyme or protein to the signal, and the sample matrix applied.

Another possibility is continuous-flow analysis with injection of enzymes or proteins into a stream of proteins or enzymes, respectively. In the first case, a continuous flow of enzymes can be infused to a mass spectrometer. In this way, the contribution of the enzyme to the signal can be observed.[20] When the signal remains stable, *i.e.* no significant autolysis is observed[21] and the spray remains stable, the protein of interest can be injected into the enzyme flow. To maintain a certain reaction time, a reaction coil can be introduced into the system. In the second case, the protein can be continuously infused and different samples containing enzymes can be injected into the flow. In such a relatively simple assay, the reaction of several types of enzymes can be studied in a medium- to high-throughput format. Also in this approach, it is very important to perform control experiments and study ionization suppression effects, *e.g.* by co-infusing and monitoring an ionization suppression indicator compound such as a related compound, *e.g.* a peptide not present in the mixture, with the carrier solution.[22,23]

Although the latter methods generally allow for fast screening of reactions at different conditions with relatively low instrumental setup complexity, these approaches suffer in most cases from incompatibility issues with the ESI-MS analysis and ionization suppression effects due to coeluting peptides. Also, direct infusion or analysis of digestion products without separation has another main disadvantage, *i.e.* the lack of differentiation possibilities between closely related proteins by means of their coeluting peptides, *e.g.* in the case of a protein drug and its metabolites. In order to address this issue, some separation step may have to be incorporated. Like in an approach using IMERs, pre- or postseparation digestion strategies can be applied to address this issue without the need to immobilize enzymes.

A highly hyphenated precolumn digestion approach is shown in Figure 6.3. The sample pretreatment was performed by an IAC column which was conditioned using a phosphate buffer at pH 7.4 (10 mM sodium phosphate, 134 mM sodium chloride and 3.4 mM potassium chloride). Protein sample

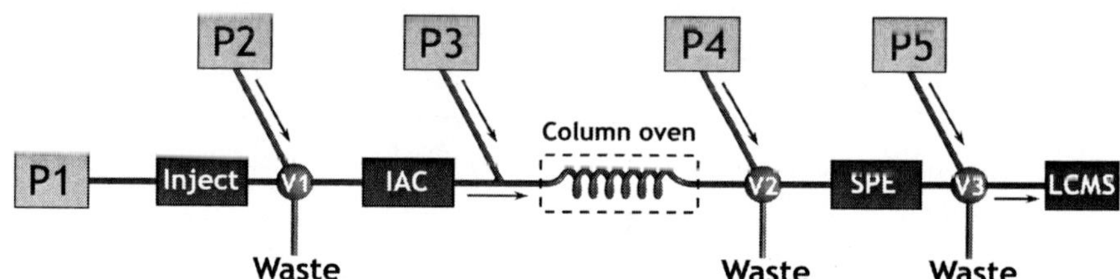

Figure 6.3 Setup of a pre-column digestion approach. Protein is injected on to the IAC column by pump 1 (P1). Pump 2 (P2) elutes the protein by switching valve 1 (V1). Pump 3 (P3) adds protease in solution. The digested product is captured on a solid phase extraction (SPE) cartridge by switching valve 2 (V2). Pump 4 (P4) desalinates the trapped peptides and elutes the peptides over the LC-MS system by switching valve 3 (V3). Pump 5 (P5) re-equilibrates the LC-MS system afterwards, while the next sample can be cleaned-up by the IAC column. (Adapted with permission from Hoos *et al.*, *J. Chromatogr., B.*, 2007, **859**, 147–156; Copyright 2007 Elsevier B.V.).

mixtures or plasma samples were injected and the sample was delivered to the IAC by pump 1 (P1). After a flushing step, the proteins of interest were eluted by switching valve 1 (V1) to pump 2 (P2) which delivered an acidic solution of glycine-HCl at pH 2.6. The protein fraction of interest was thereby delivered to the reaction capillary. Pump 3 (P3) added the enzyme in a phosphate buffer at pH 10.8, which achieved compatibility between the required acidic elution conditions of the IAC with the requirement for neutral pH during the digestion using pronase (EC 3.4.24.4) in the reaction capillary. Then, continuous flow digestion could take place under suitable temperature conditions regulated by the column oven. After the digestion, peptides were captured on an SPE cartridge, and were (after switching V2) then desalted by a flow of low organic modifier (0.5% acetonitrile) containing solution (P4). After switching V3, the desalted peptides could be separated using a gradient to a high organic modifier solution and subsequently analysed by ESI-MS^n.

The most important part in such a setup is the precolumn digestion approach in combination with the use of an enzyme reactor which can be used without the need to fabricate, validate and characterize an IMER. Furthermore, the enzyme species can be varied rather easily by infusing a different enzyme solution by pump 3. Furthermore, the sample pretreatment strategy may be varied, *e.g.* by introducing another separation dimension without the worry that the enzyme reactor will be damaged by the back-pressure of the column, which can be a major concern when using IMERs.

An example of a postcolumn digestion approach employing a continuous-flow reactor can be seen in Figure 6.4. Here, a reaction capillary is connected to the ESI-MS whereby pepsin is continuously infused by pump 3. The enzyme is delivered in pure (neutral) water, in which the enzyme is almost inactive, preventing the formation of autolysis products before the enzyme is delivered to the reaction capillary. The reaction is started only at the moment that the enzymes are mixed in the mixing union, where an acidic pH is maintained using diluted formic acid solution which is delivered by pump 2. With the reaction capillary and pumps 2 and 3, it is in principle possible to operate the in-solution digestion reaction with relatively high throughput. Altogether, the proteins could be digested in only 30 seconds and subsequently analysed. However, this particular setup was developed to analyse a protein mix. To separate a mixture

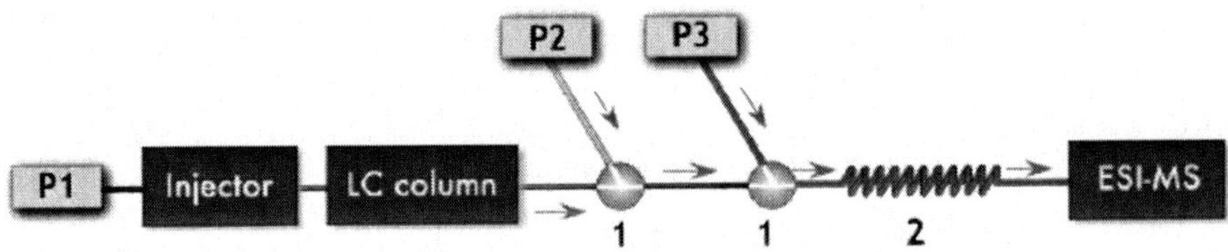

Figure 6.4   Setup of postcolumn digestion. P1, gradient pump; P2, infusion pump for pH adjustment; P3, infusion pump for enzyme addition; 1, mixing infusion; 2, fused silica reactor (Adapted with permission from Bruyneel *et al.*, *Anal. Chem.*, 2007, **79**, 1591; Copyright 2007 ACS Publications).

of six model proteins, a C-18 RPLC column was attached as a sample preparation method. In the resulting postcolumn digestion setup, a gradient was run towards high organic modifier content. However, the activity of the digesting enzyme, pepsin (E.C. 3.4.23.1) in this case, is reduced at organic modifier concentration higher than 40% acetonitrile (ACN). To overcome this incompatibility, the postcolumn infusion of aqueous formic acid diluted the column effluent containing up to 60% ACN.

In the in-solution digestion approach, digestion conditions can be varied in a rather rigorous manner. Optimum conditions may vary over a very wide range.[24,25] Especially when fresh enzymes are used continuously, parameters like temperature or pH can be varied over a very wide range to find optimum conditions, regardless of reusability of enzymes. Another factor is, of course, the digestion time, which can be varied by the reactor dimensions and the flow rate in the reactor. When optimizing the conditions, one has only to maintain the stability of the enzyme reactivity which is infused to the enzyme reactor. Recently, in-solution digestion approaches have been published presenting automated analysis of proteins, including denaturation, digestion and separation in the order of minutes,[26,27] which emphasizes the need for fast and efficient protein analysis by means of peptides without the need for manual sample handling.

6.6   Notes and Hints

Hyphenated on-line digestion can be performed using in-solution digestion and using IMERs. Although the use of IMERs may be favoured, some enzymes are not commercially available in form of an IMER. Here, it is possible to perform off-line or on-line in-solution digestion in a precolumn format with high ESI-MS compatibility or in a postcolumn digestion format keeping in mind possible ESI-MS incompatibility issues.

In the postcolumn digestion mode, the elution time of intact proteins can be monitored which allows chromatographic differentiation based on the retention time of the intact proteins by means of peptide masses present after the digestion. This may be a useful alternative to the generally used peptide mapping procedures, where the intact protein is digested and some differences may be very hard to monitor. On the other hand, peptides derived from one protein can coelute with several other peptides or proteins. This can cause ionization suppression effects hampering quantitative analysis or even hindering the analysis and identification of low abundance peptides. This demonstrates the need to perform fundamental control experiments when applying these kinds of approaches.

For some enzyme reactions, the presence of cofactors and/or high amounts of salts is required for the reaction to occur. This complicates or hinders the analysis using ESI-MS because of ionization suppression effects or multiple adduct formation (*e.g.* formation of sodium or potassium adducts and/or ions formed after H^{+}-alkali^{+}-exchange) in the ionization procedure. In this case, precolumn digestion may be favoured over direct infusion of the reaction mixture.

Although digestion using IMERs can be introduced between different analytical techniques and instruments, hyphenation of enzymatic reactions with analytical techniques can result in compatibility issues. For example, whereas RPLC of proteins is frequently performed at low pH, direct infusion of the column effluent to a trypsin IMER suffers from incompatibility issues since digestion with trypsin is performed at a neutral or slightly alkaline pH. ESI-MS analysis of the resulting peptides is preferably performed at low pH again. Because many enzyme reaction conditions differ significantly from the conditions used for ESI-MS, such compatibility issues need to be kept in mind. This can be done by carefully choosing the enzymes and/or conditions or adapting the circumstances before introducing the molecules to a certain process, *e.g.* by using a make-up flow of water to regulate the total organic modifier content in a postcolumn digestion approach, or acidic or basic buffers to compensate for a required pH value needed for a previous step.

When using hyphenated analysis systems, it is important to fully understand the separate processes in the system. This is important to enable trouble-shooting, or if it becomes necessary to alter or replace chemicals, system parts or whole instruments.

When developing a highly hyphenated system, it is most appropriate to start from the detector side of the system towards the front end, to be able to monitor the performance of the different analytical units in the on-line system, such as RPLC, IAC and/or on-line digestion.

One or more valves can easily be employed and they are (in most cases) controllable directly in the software or *via* a contact closure command.

The choice between an IMER and in-flow digestion will mainly depend on enzyme stability and availability, on the required enzyme concentration and its contribution to the resulting (LC-) ESI-MS signal. IMERs offer analysis without significant enzyme background and analysis with low enzyme consumption once the IMER has been fabricated and characterized, although the analyst needs to monitor the enzyme activity continuously. On the other hand, in-solution digestion offers the possibility of evaluating several enzymes or enzyme mixtures without concern of enzyme stability during the process, but comes with a risk of enzyme background, and the enzyme cannot be re-used after the analysis.

- ⓘ It is important to closely monitor the conditions in a digestion reaction chamber. This can be done by using detector cells, *e.g.* those available with Äkta systems (GE Healthcare, Buckinghamshire, UK)
- ⓘ *Temperature:* If using temperatures different from ambient or if it is a critical value, use a column oven with flow preheater to control the reaction temperature
- ⓘ Measure the temperature before and after the column
- ⓘ *pH:* Especially when using different pH values for processes in a hyphenated system, these need to be controlled carefully. This can be

done by using a pH (micro) detector cell (check the pH at the inlet and at the outlet of the reaction chamber or column)

- ⓘ *Other:* If necessary, critical or helpful, monitor the salt content using a conductivity cell

Evaluate the turnover rate of enzyme reactions. This can be done by varying conditions (*e.g.* by changing the pH, temperature or salt content). When optimum conditions are found for a certain reaction, the initial protein mass is present at a minimum level or is not detectable and the peptide's intensity is at a maximum level. As a control experiment, an inactivated enzyme reactor can be applied, the enzyme can be left out (if an in-solution digestion approach is employed), or the conditions can be changed to suboptimal, *i.e.* by cooling down the enzyme reactor, which can be done using a column oven.

Control experiments are a very important part of both development and routine application of (on-line) protein digestion procedures. Before evaluation of a peptide map can be achieved, additional experiments can be performed. To control the stability of the protein of interest under given conditions, a sample has to be analysed with the same conditions as the previous sample, except for addition of the enzyme. Leaving out the enzyme yields direct information on the contribution of the protein or its impurities to the peptide map. In the ideal case, only a protein ion envelope is observed in ESI-MS. Finally, the contribution of the enzyme has to be studied since active enzymes present in a sample at a significant concentration can result in autolysis. This can be done by repeating the first experiment, except that the protein which should be digested is omitted. Together with the result of the on-line digestion, these three experiments yield the required information on the proteolytic reaction under given conditions. Off-line digestion can be automated by the utilization of liquid sample handlers.

References

1. W. J. Henzel, C. Watanabe and J. T. Stults, *J. Am. Soc. Mass Spectr.*, 2003, **14**, 931.
2. K. F. Medzihradszky and A. L. Burlingame, *Methods Enzymol.*, 2005, **405**, 50.
3. S. Swatkoski, S. Russell, N. Edwards and C. Fenselau, *Anal. Chem.*, 2006, **79**, 654.
4. J. Křenková and F. Foret, *Electrophoresis*, 2004, **25**, 3550.
5. C. Temporini, E. Perani, E. Calleri, L. Dolcini, D. Lubda, G. Caccialanza and G. Massolini, *Anal. Chem.*, 2006, **79**, 355.
6. L. N. Amankwa and W. G. Kuhr, *Anal. Chem.*, 1992, **64**, 1610.
7. G. Massolini and E. Calleri, *J. Sep. Sci.*, 2005, **28**, 7.
8. J. Duan, L. Sun, Z. Liang, J. Zhang, H. Wang, L. Zhang, W. Zhang and Y. Zhang, *J. Chromatogr., A*, 2006, **1106**, 165.
9. P. J. Worsfold, *Pure Appl. Chem.*, 1995, **67**, 3.
10. B. Krajewska, *Enzyme Microb. Technol.*, 2004, **35**, 126.

11. A. M. Girelli and E. Mattei, *J. Chromatogr., B*, 2005, **819**, 3.
12. J. Carol, M. C. J. K. Gorseling, C. F. de Jong, H. Lingeman, C. E. Kientz, B. L. M. van Baar and H. Irth, *Anal. Biochem.*, 2005, **346**, 150.
13. M. L. Nielsen, M. M. Savitski, F. Kjeldsen and R. A. Zubarev, *Anal. Chem.*, 2004, **76**, 5872.
14. G. L. Corthals, R. Aebersold, D. R. Goodlett and A. L. Burlingame, *Methods Enzymol.*, 2005, **405**, 66.
15. N. I. Govorukhina, T. H. Reijmers, S. O. Nyangoma, A. G. J. van der Zee, R. C. Jansen and R. Bischoff, *J. Chromatogr., A*, 2006, **1120**, 142.
16. W. Hoesel, J. Gross, R. Moller, B. Kanne, A. Wessner, G. Müller, A. Müller, E. Gromnica-Ihle, M. Fromme, S. Bischoff and A. Haselbeck, *J. Immunol. Methods*, 2004, **294**, 101.
17. N. Kinoshita, S. Suzuki, Y. Matsuda and N. Taniguchi, *Clin. Chim. Acta*, 1989, **179**, 143.
18. J. Ma, J. Liu, L. Sun, L. Gao, Z. Liang, L. Zhang and Y. Zhang, *Anal. Chem.*, 2009, **81**, 6534.
19. J. S. Hoos, M. C. Damsten, J. S. B. de Vlieger, J. N. M. Commandeur, N. P. E. Vermeulen, W. M. A. Niessen, H. Lingeman and H. Irth, *J. Chromatogr., B*, 2007, **859**, 147.
20. A. R. de Boer, T. Letzel, D. A. van Elswijk, H. Lingeman, W. M. A. Niessen and H. Irth, *Anal. Chem.*, 2004, **76**, 3155.
21. X.-F. Li, X. Nie and J.-G. Tang, *Biochem. Biophys. Res. Commun.*, 1998, **250**, 235.
22. T. M. Annesley, *Clin Chem.*, 2003, **49**, 1041.
23. C. F. de Jong, R. J. E. Derks, B. Bruyneel, W. Niessen and H. Irth, *J. Chromatogr., A*, 2006, **1112**, 303.
24. S. C. Cheison, M. Schmitt, E. Leeb, T. Letzel and U. Kulozik, *Food Chem.*, 2010, **121**, 457–467.
25. S. C. Cheison, E. Leeb, T. Letzel and U. Kulozik, *Food Chem.*, 2011, **125**, 121.
26. D. López-Ferrer, K. Petritis, N. M. Lourette, B. Clowers, K. K. Hixson, T. Heibeck, D. C. Prior, L. Pasă-Tolić, D. G. Camp, M. E. Belov and R. D. Smith, *Anal. Chem.*, 2008, **80**, 8930.
27. J. Sproß and A. Sinz, *Anal. Chem.*, 2010, **82**, 1434.

CHAPTER 7

Bioinformatic Tools for the LC-MS/MS Analysis of Proteins and Peptides

CHRISTIAN WEBHOFER[1] AND MICHAEL SCHRADER[2]

[1] Max Planck Institute of Psychiatry, Proteomics and Biomarkers, Professor Dr. Christoph W. Turck, Kraepelinstrasse 2-10, 80804 Munich, Germany; [2] Weihenstephan-Triesdorf University of Applied Sciences, Department of Biotechnology and Bioinformatics, 85350 Freising, Germany

7.1 Introduction

7.1.1 General

Tandem mass spectrometry coupled to liquid chromatography became a major analytical method in protein and peptide analysis after the introduction of soft ionization methods.[1] Another essential requirement was the development of software tools and biological databases. This issue did not attract the same awareness, but sophisticated tools are needed to interpret the wealth of complex information represented within mass spectra. A molecular mass might be interpreted manually, but fragmentation patterns of biomolecules by MS/MS (tandem mass spectrometry) and quantification of data from LC-MS/MS (combination of separation by liquid chromatography and detection by MS/MS) must be automated to a substantial extent.

In bottom up proteomics, MS/MS is not directly applied on proteins. Proteins are cut into peptides by specific proteases—most often trypsin is used. This leads to more complexity, but data quality and separation efficiency

RSC Chromatography Monographs No. 15
Protein and Peptide Analysis by LC-MS: Experimental Strategies
Edited by Thomas Letzel
© The Royal Society of Chemistry 2011
Published by the Royal Society of Chemistry, www.rsc.org

improve much more. Therefore, protein analysis is usually performed *via* peptide analysis. One protein is typically digested into 10–100 tryptic peptides, depending on protein molecular weight and amino acid sequence. The sheer amount of data has to be automatically reduced to the information essential for sequence identification. However, due to technical and biological variability, an experienced user is still required, not only to generate the data, but also to guide the data analysis and finally interpret the information content.

Here, a basic introduction will be given to the principal functionality of software tools for MS/MS analyses for peptide identification. This exemplification is addressed to readers with no former experience in MS/MS. Moreover, as an example we illustrate an automated LC-MS/MS workflow for protein identification and quantification by *in vivo* ^{15}N metabolic labelling.

7.1.2 Protein and Peptide Sequence Analysis by MS/MS

Presently, most peptide and protein sequence information is generated by LC-MS methods combined with database search algorithms. This reflects the high-throughput capabilities of the technique as compared to earlier methods such as Edman sequencing or amino acid analysis. Alternatively, in off-line methods sequences of proteins can also be identified after initial purification followed by trypsin digestion to produce characteristic peptide mass fingerprints by MALDI-MS.[2] Sequence identification is then performed by comparison with *in silico* digests of large protein databases.

Mixtures of proteins cannot be analysed accordingly. Therefore, in many cases another type of sequence identification is needed. A common solution is to additionally separate tryptic peptides by liquid chromatography (LC). Peptides eluting in these complex mixtures will be subjected to fragmentation analysis by MS/MS.[3] Modern MS instrumentation allows for fast switching between full scan and MS/MS analysis. Within an LC run, it is thus possible to determine the molecular masses of separated peptides as well as their corresponding MS/MS spectra. In the molecular mass range of tryptic peptides, all standard hybrid ESI-MS/MS instrumentation (*e.g.* ion trap, quadrupole–TOF, Orbitrap) can be used.[4] Peptides generated by other enzymes or top down analyses of intact native peptides can be processed in the same way. Most current spectrometers provide a sufficiently high resolution.[5,6]

7.1.3 Software Tools for Peptide Sequence Interpretation by MS/MS

A single mass spectrometer can easily generate thousands of MS/MS spectra per LC-MS run. Many software tools are currently available for correct and rapid assignment of peptide sequences by database comparison.[7] Their original development dates back more than a decade, resulting in robust software which in many cases is even freely available on the internet.[8] A useful list of free software for interpretation of MS/MS data can be found at www.ms-utils.org.

Table 7.1 Alphabetical list of most prominent free software tools allowing peptide sequence interpretation by MS/MS data by web access

Name of software	Website	Organization
Mascot	matrixscience.com	Matrix Science, London
OMSSA	pubchem.ncbi.nlm.nih.gov//omssa/	NIST, Bethesda
Phenyx	phenyx.vital-it.ch/pwi	Geneva Bioinformatics
Protein Prospector (MS-Seq)	prospector.ucsf.edu/	Univ. of California, San Francisco

Table 7.2 Alphabetical list of widely used commercial software tools for peptide sequence interpretation by MS/MS delivered by mass spectrometry vendors (usually integrated in software distribution)

Name of software	Vendor	Website
BioTools	Bruker Daltonics	www.bdal.com
MassLynx	Waters	www.waters.com
SEQUEST	Thermo Scientific	www.thermo.com
SpectrumMill	Agilent	www.chem.agilent.com

On the other hand, every mass spectrometer usually is delivered with vendor software allowing this kind of data interpretation. As all software packages rely basically on similar principles, they lead to comparable results. However, it should be mentioned that different software may result in different protein identifications due to distinct scoring of peptide matches and protein assignments.[9] Some of the commonly used free and commercial tools are listed in Tables 7.1 and 7.2 respectively.

All of these software tools rely on comparisons of MS/MS peak lists against theoretical fragmentation patterns obtained from protein sequence databases.[9] With the enormous growth of publicly available sequence data from genome sequencing projects, only protein sequences of rare organisms are not publicly available in databases. Even in those cases, at least homologous sequences will be found in many attempts. Otherwise, so-called *de novo* interpretation is necessary, which is much more complicated and needs other software tools.[4]

Mascot is a commercial software tool, but since its launch it has been freely available *via* a web interface (Table 7.1).[8,10] It is used here as an example of a simple MS/MS analysis. The following automated MS/MS analyses for quantification are based on SEQUEST.[11]

7.1.4 Quantification by LC-MS/MS after Isotopic Labelling

Over the past decade, a series of experimental strategies for MS-based quantitative proteomics and corresponding computational methodology for the processing of resulting data have been generated.[12] These allow for the analysis of data from complex experiments by combining several processing steps.

An overview of the main quantification principles and available software solutions for the analysis of data generated by liquid chromatography coupled to mass spectrometry (LC-MS) can be found in Mueller *et al.*[13] or Vaudel *et al.*[14] A quantitation workflow based on *stable isotope labelling by amino acids in cell culture* (SILAC[15]) is explicitly described in Chapter 8 of this book. Further strategies are described in Chapter 1.

Here, we focus on quantification by LC-MS/MS after *in vivo* ^{15}N metabolic labelling. This is a powerful tool for accurate high-throughput protein quantification. Labelling of proteins is achieved by introducing fully ^{15}N-labelled amino acids during protein synthesis. Light (^{14}N) and heavy (^{15}N) protein samples are combined prior to sample preparation in order to increase quantitation accuracy and enable measurement of corresponding peptide peaks in a single MS spectrum.

^{14}N and ^{15}N protein isoforms behave identically during sample preparation and mass spectrometric analysis, *e.g.* in terms of chromatographic retention time and ionization efficiency. However, heavy ^{15}N isoforms are distinguishable from its light ^{14}N counterparts through a right shift (higher m/z) in the mass spectrum and relative quantification is achieved by comparing ^{14}N and ^{15}N isotopologue intensities. ^{15}N sample intensities are then used as internal standard (IS) for the indirect comparison of two unlabelled ^{14}N samples A and B, *e.g.* healthy *versus* control or treated *versus* untreated. The following equation for indirect comparison is applied:

$$\text{sample A}/^{15}\text{N IS} : \text{sample B}/^{15}\text{N IS} = \text{sample A} : \text{sample B}.$$

Here we present an exemplary study of relative protein quantification using *in vivo* ^{15}N metabolic labelling of mice, including software tools for data analysis and interpretation.

7.2 Materials

7.2.1 Peptides

- Protein sources: Cell lysate from *E. coli BL21star* (Invitrogen, San Diego, CA, USA), protein extract from hippocampus of DBA/2Ola mice (Harlan Winkelmann, Borchen, Germany)
- Enzyme: Trypsin NB premium grade (Serva Electrophoresis, Heidelberg, Germany)

7.2.2 LC-MS

- HPLC: NanoLC-2D (Eksigent, Dublin, CA, USA)
- LTQ Orbitrap mass spectrometer (Thermo Fisher Scientific, Bremen, Germany)
- Xcalibur 2.0.6 (Thermo Fisher Scientific)

7.2.3 Protein Identification by MS/MS

- Mascot (Matrix Science, Boston, MA, USA) for exemplary MS/MS data analysis
- Bioworks (Thermo Fisher Scientific), based on SEQUEST
- DTASelect v1.9,[16] for filtering results from database searches

7.2.4 Protein Quantification

- The Atomizer[17]
- RelEx,[18] available at http://fields.scripps.edu/relex/
- ReAdW, available at http://sourceforge.net/projects/sashimi/files/
- ProRata,[19] available at http://code.google.com/p/prorata/updates/list

7.2.5 Data Analysis and Interpretation

- MetaboAnalyst,[20] http://www.metaboanalyst.ca/MetaboAnalyst/faces/Home.jsp
- DAVID Bioinformatics Resources, http://david.abcc.ncifcrf.gov/tools.jsp
- Pathway Studio 7.1 (Ariadne Genomics, Rockville, MD, USA)

7.3 Methods

7.3.1 Sample Preparation for LC-MS

- Extract proteins from mouse tissue using a protocol of choice. We suggest using a protocol by Emili and Cox that allows for simultaneous extraction of mitochondrial, nuclear, cytosolic and microsomal proteins from a given tissue.[21]
- Determine protein concentrations by the Bradford method[22] and mix ^{14}N and ^{15}N samples at equal protein amounts
  - It is optimal to perform the Bradford assay for all samples at the same time in order to minimize interexperimental variability
  - As the ^{15}N sample is further used as internal standard it is necessary to use the identical ^{15}N protein sample for all ^{14}N samples. If the protein amount derived from one animal is not sufficient, combine several extracted ^{15}N protein samples
- Load 100 µg of each ^{14}N/^{15}N mixture on SDS gel and separate proteins by gel electrophoresis. Fix and stain proteins, then destain gel
  - Optimally run all biological replicates on the same gel for identical electrophoretic conditions
  - Try to avoid extensive staining of proteins, as protein-bound Coomassie will interfere with subsequent in-gel digestion
- Perform in-gel digestion according to protocol of choice, *e.g.* as described by Rosenfeld *et al.*[23] Typically, proteins are destained, disulfide bridges are reduced and cysteine residues are alkylated using iodoacetamide.

Digest proteins with trypsin overnight and extract peptides from the gel. Vacuum dry and store peptides at $-20\,^{\circ}C$ until LC-MS/MS analysis

7.3.2   LC-MS Analysis

- Dissolve peptides in 0.1% formic acid and perform LC-MS/MS analysis. Peptides are loaded on a C-18 precolumn for on-line desalting. Peptides are separated on a C-18 analytical column by a linear gradient from 10% to 50% acetonitrile. Eluting peptides are sprayed directly into the mass spectrometer. Full spectra are acquired in an Orbitrap mass analyser at a resolution of 60 000 (at $m/z = 400$). The five most intense peaks are subjected to fragmentation in a data-dependent manner
  - ➢ Do not load the total sample, in order to avoid column overload which leads to reduced chromatographic performance. Moreover, in case of an unexpected error during LC-MS/MS the analysis sample may be re-run

7.3.3   Exemplary MS/MS Data Analysis by Database Comparison

In the following section a rather simple analysis of a single peptide is explained in detail. This exemplification mainly addresses readers with no previous experience of MS/MS analysis. This rather artificial example can be traced by the web interface of Mascot to illustrate input and output as well as the corresponding parameters of this example search algorithm. These should be kept in mind in order to understand automated analysis procedures. A more advanced reader might proceed directly with section 7.3.4.

7.3.3.1   *MS/MS Peak Lists*

- An MS/MS spectrum is always the starting point. Any typical software package will reduce this to a peak list. There is a Mascot-specific data format; however, common vendor specific formats are also applicable, as well as XML data, a common standard for mass spectrometric files
- Data files fed into search tools have a rather simple structure. They consist only of a list of all peaks of the MS/MS spectrum and the molecular mass and charge of the chosen precursor ion. About 50 to several hundred entries are typically included, whereas successful database searches might be accomplished with less information. Save the reduced dataset given in Table 7.3 as explained there
- *Longer peak lists:* The original peak list contains more than 500 entries (instead of 11). With this file the same result is achieved. Its probability is only a little higher, although many more signals can be identified (see later in Figure 7.2). In conclusion, high mass accuracy is needed, but not as many peaks as possible. Some 10–100 peaks chosen properly are usually sufficient for successful identification

Table 7.3 Exemplary peak list generated by reduction of an MS/MS of a tryptic *E. coli* peptide to the most intense peaks. Data is given as data-file format of Sequest software (a shortened format which is processable by Mascot). The first line contains information about the precursor molecule (in bold), the following lines show mono-isotopic *m/z*-values of the 11 most intense fragment ions and their respective intensities. The numbers given in the latter two rows might be copied and separated each by a space character to verify the searches explained in the following. The dataset should be saved as filename.dta

Line	Monoisotopic m/z of precursor/fragment ions	Charge number/ intensity
1	**1796.95**	**2**
2	794.45	542
3	867.28	1736
4	868.07	325
5	939.49	373
6	1003.50	957
7	1004.46	361
8	1076.53	327
9	1140.48	735
10	1141.55	516
11	1253.57	230
12	1452.64	209

7.3.3.2 Input Parameters and Output

An adequate choice of search parameters is of crucial importance. Several specific decisions have to be taken through the main interface of Mascot to allow an unambiguous and fast analysis. Depending on the experimental setup, different modifications may be applied. An overview is given in Figure 7.1.

- Choice of database:
  - ➢ Must be in accordance with the species of interest. Many of the accessible databases are based on expressed sequence tags (*i.e.* derived from fast cDNA sequencing) and may contain single sequence errors
  - ➢ SWISS-PROT, as a very well curated database, is a good choice if human proteins or those of several other model organisms are searched for. As this database is also relatively small, searches will be faster than with huge EST databases. The MS/MS peak list shown in Table 7.3 was generated from a tryptic digestion of *E. coli* proteins (an organism with relatively low complexity of protein content)
- Choice of enzyme and miscleavages:
  - ➢ Enter the type of enzyme used to cleave the protein (here: trypsin)
  - ➢ Determine the number of possible miscleavages of the enzyme:
    For trypsin with properly chosen experimental conditions this parameter, should not exceed 1 or very seldom 2

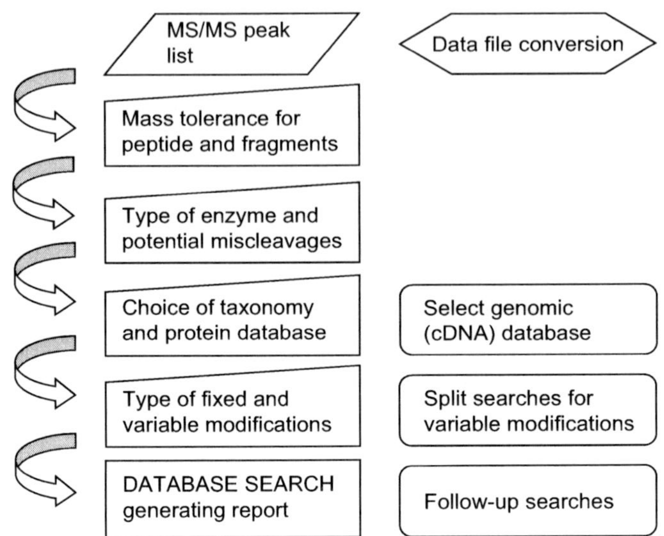

Figure 7.1   Workflow of MS/MS analysis (left) and potential additional steps (right) summarizing decisions for input parameters. Please note that splitting of searches needs some experience and interpretation might become rather difficult.[24]

> For unspecific enzymes like pepsin or tightly structured proteins, this value could be much higher. It is restricted to an upper value of 9 as anything higher leads to perturbations in sequence variations thus exceeding search times and decreasing significance

> Exemplary search: After uploading the data-file as well as choosing trypsin with one miscleavage site and SWISS-PROT database, the search result is the sequence TELHSALKSSNLNLIR (using the single-letter amino acids code). There is one tryptic cleavage site within this sequence. Its probability is not convincingly significant (indicated by the greenish shade in the figure shown by Mascot). In Mascot, the ion score for an MS/MS match is based on the calculated probability, P, that the observed match between the experimental data and the database sequence is a random event. Moreover, the molecular mass deviates by 0.95 Da (600 ppm), which is far too high for modern mass spectrometry. That example shows the importance of high mass accuracy to decrease false-positive identifications. Mass tolerance values are usually set up-front (see Figure 7.1). With rigid mass tolerance settings this candidate would be omitted

- Choice of potential modifications:
  > Potential peptide modifications might be due to native, post-translational modification like phosphorylation, glycosylation or N-terminal pyroglutamic acid. As these modifications are not included in the searchable part of protein and DNA-based databases, corresponding shifts in molecular mass have to be corrected by the

search program. Other modifications might stem from sample handling with chemicals that might react with functional groups of the protein (*e.g.* alkylation of cysteines) or simply oxidation of amino acid residues (especially Met and Trp) before or even during the measurement. Both types of modifications are mostly specific and thus related to a certain type of amino acid residue which is given in parentheses in the menu

> *Example search:* The initial sequence determination was incorrect. In our sample case oxidation of a Met residue took place in the peptide chain. Set 'Oxidation (M)' as 'fixed' modification. The following search delivers the same wrong result. Fixed modifications are used only if the chemistry (or biology) is known to be very specific and affect every amino acid residue

> Now set 'Oxidation (M)' as 'variable' modification allowing an optional oxidation of methionine. A clear hit appears, with the sequence MVVTLIHPIAMDDGLR. It contains two Met residues, one of them oxidized, which increases the molecular mass substantially by 15.995 Da. Even one unknown modification can disturb the database search substantially, leading to no result at all. The final search results are given in Figure 7.2

7.3.4 Automated MS/MS Data Analysis for Quantification

7.3.4.1 Protein Identification

- MS/MS spectra are searched against a uniprot.MOUSE database using the search engine *Sequest* implemented in the software *Bioworks* (Figure 7.3)
- The following search parameters are used:
  - MS precursor mass accuracy: 15 ppm
  - MS/MS fragment mass accuracy: 0.8 Da
  - Missed tryptic cleavage sites allowed per peptide: 2
  - Fixed peptide modification: Carbamidomethylation of cysteine
  - Variable peptide modification: Oxidation of methionine
  - > Depending on the experimental setup, different modifications may be applied
  - > When using lock masses[25] during MS acquisition for internal mass calibration, more stringent precursor mass accuracy may be applied to facilitate peptide identification
- For identification of MS/MS spectra derived from ^{15}N precursor ions, perform a ^{15}N database search if ^{15}N incorporation is higher than approximately 80%. Use an increment of +0.997 Da for all nitrogens per amino acid as fixed modification. Note that nitrogen introduced from iodoacetamide during Cys alkylation remains natural ^{14}N
  - > In case of incomplete ^{15}N labelling (approximately 80–95% ^{15}N incorporation) use a variable modification of −0.997 Da per arginine or lysine.

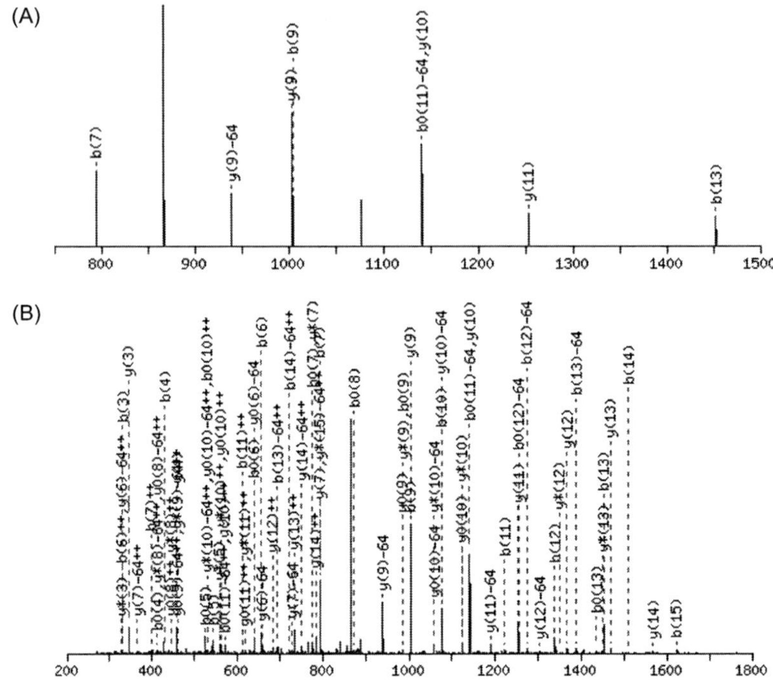

Figure 7.2 Mascot search results for MS/MS corresponding to tryptic *E. coli* peptide MVVTLIHPIAMDDGLR in which one M is oxidized. (A) Reduced dataset containing 11 fragment peaks, using 10 of which 7 were assigned; (B) original dataset of more than 500 peaks, 142 used and 77 assigned. The search in (A) already delivered the correct peptide sequence. Its probability is not much lower, although many more signals could be identified in (B). Too many peaks lead to ambiguities, thus Mascot reduced this list to 142 significant peaks by an internal algorithm.

This accounts for the '1 Da' left shift from the 'monoisotopic' ^{15}N peak to the most intense ^{15}N peak which is subjected to fragmentation.[26]

- Filter peptide identifications based on MS/MS quality criteria using *DTASelect*
    - Stringent, commonly applied criteria for *DTASelect* are: DeltaCN = 0.08 and Xcorr *versus* ChargeState = 2.7(2+), 3.5(3+) and 3.0(>3+). The filtered identification file is called dta-select-filter.txt
    - Alternatively, adjust filter criteria in order to reach a False Discovery Rate of 1% at the peptide level using a target-decoy search strategy.[27]

7.3.4.2 *Protein Quantification*

- Identified proteins are relatively quantified using the software *ProRata*. The dta-select-filter file is used as identification input and mzXML files are used for MS spectral information. Quantification parameters are set in the ProRata Config-file

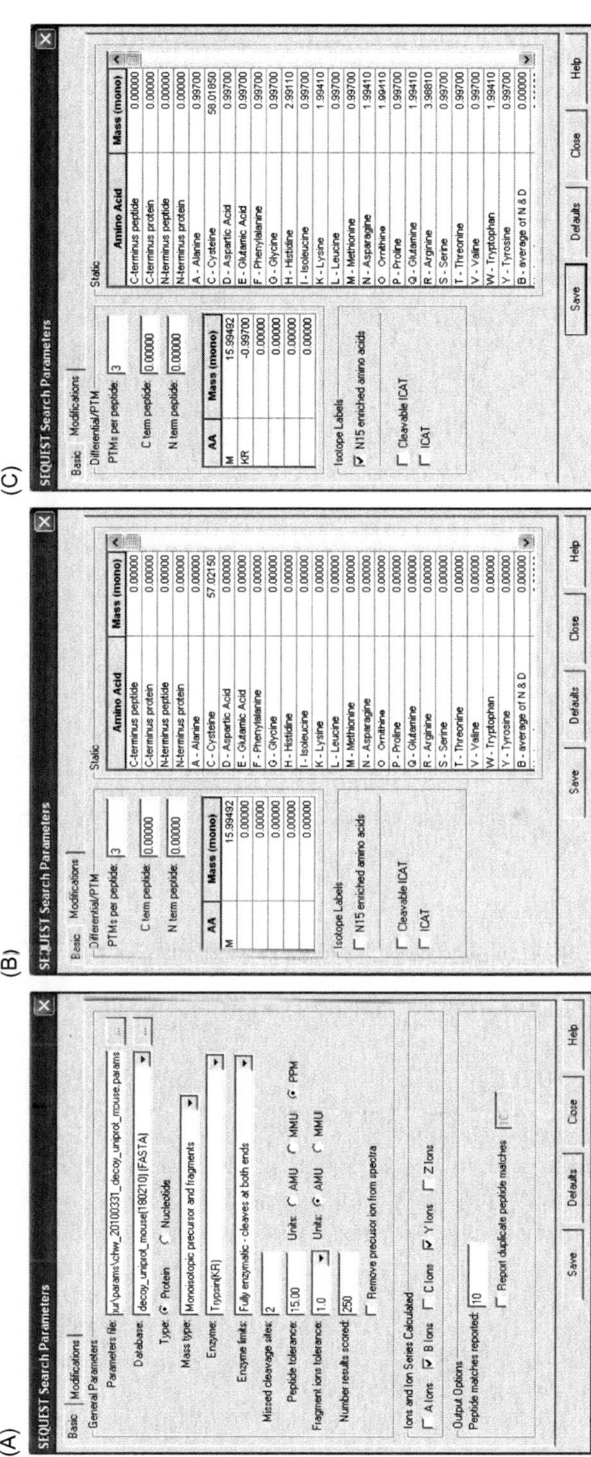

Figure 7.3   (A) General SEQUEST setting. (B) Amino acid modifications for ^{14}N database search. (C) Modifications for ^{15}N database search.

- ➢ A crucial parameter in the Config file is the ^{15}N incorporation percentage. For optimal quantification accuracy, the ^{15}N incorporation should first be determined using software tools like *The Atomizer* or *RelEx*
- For protein quantification, ^{14}N and ^{15}N peptide ion chromatograms are extracted (XIC) based on peptide retention time, ^{15}N incorporation and isotopologue intensities and transformed into a 'peak profile'. Principal component analysis of the profile is used to estimate the peptide abundance ratio and to score the results with the signal-to-noise ratio. Individual peptide information is further combined to calculate a protein abundance factor with 95% confidence interval, as depicted in the 'protein profile likelihood'. Using the ProRata-combine script profile likelihoods of different biological replicates are combined and indirect comparison *via* the ^{15}N internal standard is performed (Figure 7.4)
- The final output result file will comprise a list of quantified proteins along with information on protein abundance factors, 95% confidence intervals and the number of peptides used for quantification
  - ➢ Figure 7.5 shows an exemplar distribution of abundance factors with 95% confidence intervals

7.3.5   Identification of Significantly Different Proteins

- Significant changes may be assessed using the 95% confidence intervals. Considering proteins whose confidence interval does not contain the 1:1 ratio ($\log_2 = 0$) corresponds to a two-tailed t-test with significance level of 5%
  - ➢ Note that this kind of analysis does not correct for multiple testing and significant results should be treated with caution as it could contain a high number of false positives.[28]
- *MetaboAnalyst* may be used to identify significantly altered proteins. You may use *SAM* (*Significant Analysis of Microarrays*) instead of 95% confidence intervals as *SAM* accounts for multiple testing by adjusting False Discovery Rates. Thus the number of false-positives is reduced, whereas the number of false-negatives is still reasonable
- Depending on the number of biological replicates and quantified proteins you may perform multivariate data analysis, like *PCA (Principal Component Analysis)* or *PLS-DA (Partial Least Square–Discriminant Analysis)*
- In order to increase data robustness and confidence in the results, two different workflows of data analysis may be performed and only those proteins that are identified by both methods considered for further analysis
- For further reading on bioinformatics-based data analysis strategies we suggest Azuaje *et al.*[29]

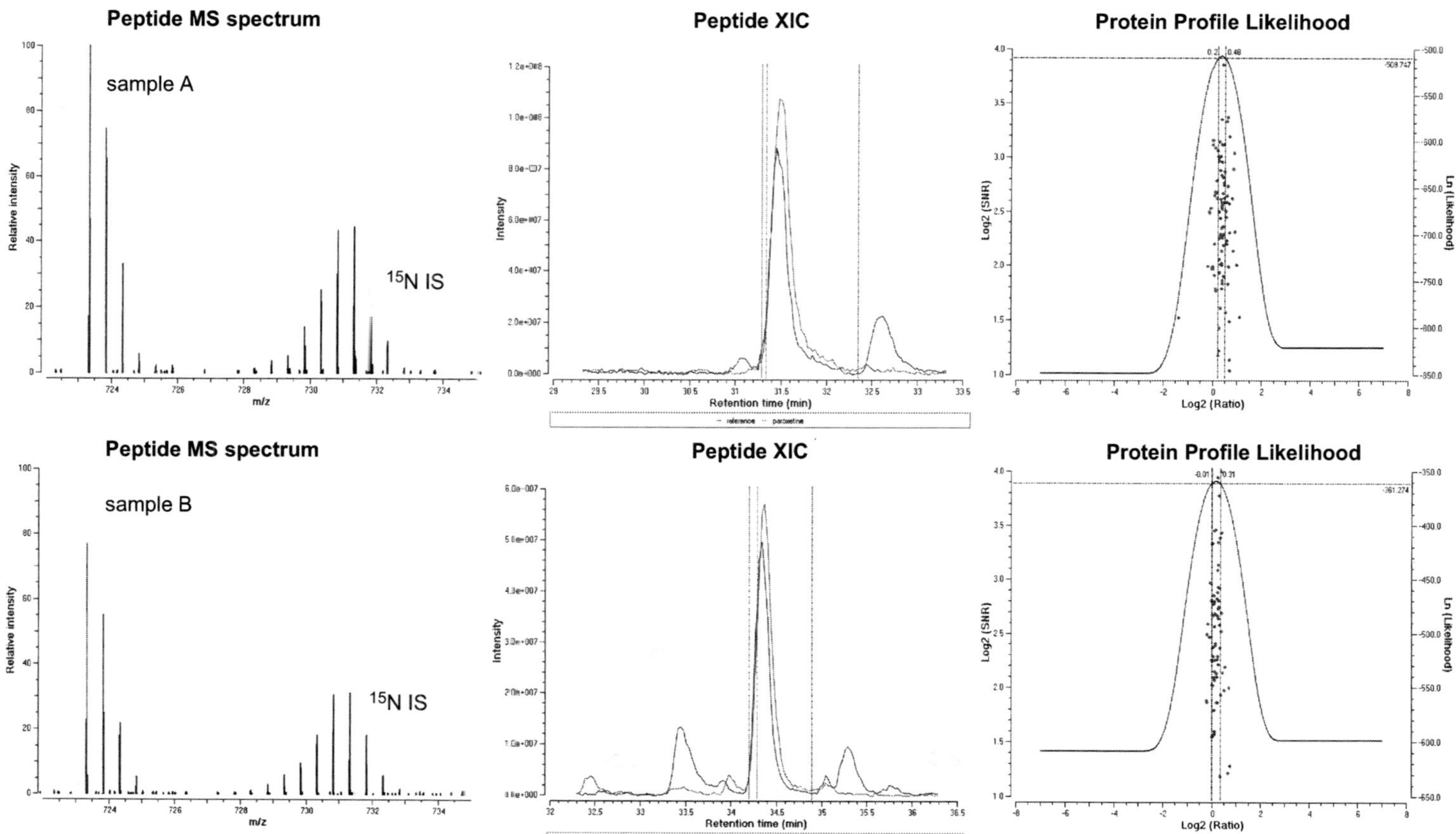

Figure 7.4 Example workflow from $^{14}N/^{15}N$ peptide mass spectrum to extracted ion chromatograms (XIC) to calculate $^{14}N/^{15}N$ protein abundance factors. With help of the protein profile likelihood profile, a 95% confidence interval for each protein abundance factor is calculated.

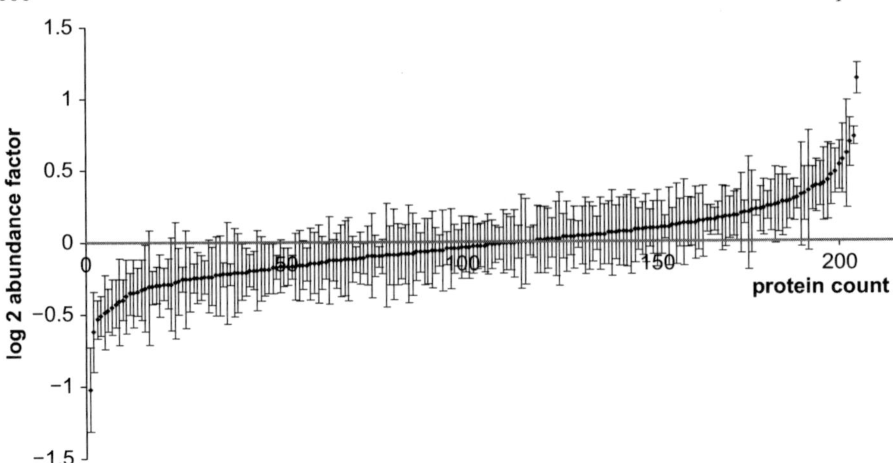

Figure 7.5   Protein abundance factors with 95% confidence intervals for about 200 representative proteins. The horizontal line represents the 1:1 ratio ($\log_2 = 0$). If error bars are clearly above or below this line, a significant alteration in protein concentration might be expected.

7.3.6   Data Interpretation

After thorough data analysis, some hits hopefully remain that fulfill all criteria for significance. If this is true, some early interpretation of biological context might be worthwhile.

- The Database for Annotation, Visualization and Integrated Discovery (DAVID) offers tools for the identification of enriched KEGG (Kyoto Encyclopedia of Genes and Genomes) or GO (Gene Ontology) classes
  - ➤ It includes a very useful description of protein properties
  - ➤ Functional annotation clustering is a nice tool to identify enriched processes based on diverse protein databases
- *Pathway Studio* is a powerful tool for the identification of affected molecular pathways
  - ➤ Interrogate your protein data for already published connections to other proteins, metabolites, cellular processes or diseases

7.4   Notes

7.4.1   MS/MS Database Searches

- Tryptic fragments of proteins deliver very good fragment spectra. Corresponding peptides are mainly in a suitable range of 500–2500 Da and contain a basic amino acid at their C-terminus facilitating the fragmentation pattern. Almost all MS/MS software tools were developed using such spectra
- What is a good MS/MS spectrum? It should contain more than at least 10 mass spectrometric signals with a signal-to-noise ratio greater than 10.

The signals should spread over the entire mass range. A few very intense signals may lower the overall quality. More than 50 significant signals do usually not improve the quality of the searches any further
- Database searches are very effective and can be automated. However, false-negatives are very common such that no appropriate hit is found with standard settings. If statistical ranking is used properly, the risk of ending up with false-positives is usually much lower (depending on actual mass accuracy)
- EST sequence data derived from fast cDNA sequencing is generated in high-throughput fashion and may include single errors in amino acid sequences. Search for homologues by allowing error-tolerant searches and validate initial hits. In contrast, sequences in SWISS-PROT or other well-curated databases are mainly correct
- Setting of modifications is crucial and cannot be determined from the database. If these are expected, they must be chosen in advance as molecular masses of respective fragments differ. An automated analysis would lead to no significant or even false-positive results. Limit the number of variable modifications to those expected.[6]
- Variable modifications lead to an increase in sequences to match and thus to both increased search times and decreased significance. Sometimes, it might thus be helpful to perform multiple, independent searches for different modification settings
- If you do not get a hit for a good fragmentation spectrum, it is worthwhile to check for the combination of sequence homologues, post-translational modifications and mass spectrometry adducts, since these are not found during a regular database search. However, with extensive knowledge of your sample and some experience, you might be able to solve these spectra

7.4.2 Quantitative Determinations by LC-MS/MS

- Internal and external standardization is necessary to reduce experimental variation, *e.g.* by using ^{15}N labelled proteins as internal standard for indirect protein quantification or lock masses for internal mass calibration during MS analysis
- Usually, peptides from the same protein ID are identified in more than one SDS gel slice. Combine peptide identifications from adjacent gel slices, as these probably derive from the same protein isoform. However, identifications from different regions of the gel should be treated individually as these may be derived from distinct protein isoforms or cleavage products
- Consideration of confidence intervals, typically set to 95%, is crucial to receive results with sufficient quantification accuracy. Automated searches in particular should be performed with rigid constraints. Do not report protein quantification ratios without a measure of accuracy
- Biological studies should always be performed using a reasonable number of biological replicates. However, as there is no consensus for the

integration of biological replicates to identify significantly altered protein levels, make some effort to pursue an optimized data analysis strategy for your dataset
- Never start biological interpretation of intermediate results! Validate your hits carefully and take care of proper statistical evaluation. It is usually better to reject a hit instead of following a false positive. In almost all cases the wealth of biological literature will give several seemingly relevant hints and you will waste a lot of time with insignificant results. Focus on the robust candidates you have, or even collect additional data

References

1. R. Aebersold and M. Mann, *Nature*, 2003, **422**, 198.
2. R. Cramer, in *Proteomics*, ed. J. Reinders and A. Sickmann, Humana Press, Totowa, NJ, 2009, p. 85.
3. W. Staudenmann and P. James, in *Proteome Research: Mass Spectrometry*, ed. P. James, Springer, Berlin, 2001, p. 143.
4. B. Lu, T. Xu, S. K. Park and J. R. Yates 3rd, in *Proteomics*, ed. J. Reinders and A. Sickmann, Humana Press, Totowa, NJ, 2009, p. 261.
5. T. Möhring, M. Kellmann, M. Jürgens and M. Schrader, *J. Mass. Spectrom.*, 2005, **40**, 214.
6. K. Sasaki, Y. Satomi, T. Takao and N. Minamino, *Mol. Cell. Proteomics*, 2009, **8**, 1638.
7. P. Hernandez, M. Müller and R. D. Appel, *Mass Spectrom. Rev.*, 2006, **25**, 235.
8. F. Lisacek, *Proteomics*, 2006, **6**, S22.
9. A. I. Nesvizhskii, in *Mass Spectrometry Data Analysis in Proteomics*, ed. R. Matthiesen, Humana Press, Totowa, NJ, 2007, p. 87.
10. D. N. Perkins, D. J. Pappin, D. M. Creasy and J. S. Cottrell, *Electrophoresis*, 1999, **20**, 3551.
11. J. K. Eng, A. L. McCormack and J. R. Yates 3rd, *J. Am. Soc. Mass Spectrom.*, 1994, **5**, 976.
12. M. Bantscheff, M. Schirle, G. Sweetman, J. Rick and B. Kuster, *Anal. Bioanal. Chem.*, 2007, **389**, 1017.
13. L. N. Mueller, M. Y. Brusniak, D. R. Mani and R. Aebersold, *J. Proteome Res.*, 2008, **7**, 51.
14. M. Vaudel, A. Sickmann and L. Martens, *Proteomics*, 2010, **10**, 650.
15. S. E. Ong, B. Blagoev, I. Kratchmarova, D. B. Kristensen, H. Steen, A. Pandey and M. Mann, *Mol. Cell. Proteomics*, 2002, **5**, 376.
16. D. L. Tabb, W. H. McDonald and J. R. Yates 3rd, *J. Proteome Res.*, 2002, **1**, 21.
17. M. J. MacCoss, C. C. Wu, D. E. Matthews and J. R. Yates 3rd, *Anal. Chem.*, 2005, **77**, 7646.
18. M. J. MacCoss, C. C. Wu, H. Liu, R. Sadygov and J. R. Yates 3rd, *Anal. Chem.*, 2003, **75**, 6912.

19. C. Pan, G. Kora, W. H. McDonald, D. L. Tabb, N. C. VerBerkmoes, G. B. Hurst, D. A. Pelletier, N. F. Samatova and R. L. Hettich, *Anal. Chem.*, 2006, **78**, 7121.
20. J. Xia, N. Psychogios, N. Young and D. A. Wishart, *Nucleic Acids Res.*, 2009, **37**, W652.
21. B. Cox and A. Emili, *Nat. Protoc.*, 2006, **1**, 1872.
22. M. M. Bradford, *Anal. Biochem.*, 1976, **72**, 248.
23. J. Rosenfeld, J. Capdevielle, J. C. Guillemot and P. Ferrara, *Anal. Biochem.*, 1992, **203**, 173.
24. R. Tharakan, N. Edwards and D. R. Graham, *Proteomics*, 2010, **10**, 1160.
25. J. V. Olsen, L. M. de Godoy, G. Li, B. Macek, P. Mortensen, R. Pesch, A. Makarov, O. Lange, S. Horning and M. Mann, *Mol. Cell. Proteomics*, 2005, **4**, 2010.
26. Y. Zhang, C. Webhofer, S. Reckow, M. D. Filiou, G. Maccarrone and C. W. Turck, *Proteomics*, 2009, **9**, 4265.
27. J. E. Elias and S. P. Gygi, *Nat. Methods*, 2007, **4**, 207.
28. T. K. Rice, N. J. Schork and D. C. Rao, *Adv. Genet.*, 2008, **60**, 293.
29. F. Azuaje, ed., *Bioinformatics and Biomarker Discovery: 'Omic' Data Analysis for Personalized Medicine*, Wiley-Blackwell, 2010.

Quantitative LC-MS of Proteins

GABRIELE STÖHR[1] AND ANDREAS TEBBE[2]

[1] Max Planck Institute of Biochemistry, Department of Proteomics and Signal Transduction, Am Klopferspitz 18, 82152 Martinsried, Germany; [2] KINAXO Biotechnologies, Am Klopferspitz 19a, 82152 Martinsried, Germany

8.1   Introduction

Proteins are the key players in almost all cellular processes. Consequently, methods are needed that allow the system-wide analysis of proteins for the description of cellular events. Liquid chromatography coupled to mass spectrometry (LC-MS) has become the method of choice to analyse complex protein samples as it permits the automated sequencing of thousands of peptides in a reasonable time frame.[1] Despite this ability the simple determination of protein inventories is in most cases not sufficient to address complex biological questions. An additional dimension is clearly needed in the mass spectrometric analysis that enables the quantitative comparison of different samples.

Generally, mass spectrometry itself is not a quantitative technique as different molecules show different responses in the mass spectrometer due to *e.g.* different ionization efficiencies of different molecules (see also Chapter 2). Consequently, new tools had to be introduced to turn mass spectrometry based protein analysis into a quantitative technology. Today, a multitude of techniques are at hand for the absolute and relative quantification of proteins between different samples.[2–4] Mostly, all of those approaches rely on the incorporation of different forms of stable isotopes into the proteins or peptides or the use of isotopically labelled reference peptides (see also Chapter 1). The introduction of the stable isotopes

RSC Chromatography Monographs No. 15
Protein and Peptide Analysis by LC-MS: Experimental Strategies
Edited by Thomas Letzel

Published by the Royal Society of Chemistry, www.rsc.org

can be achieved by different means such as chemical, enzymatic or metabolic labelling.[5–7] This labelling step gives rise to protein samples being distinguishable in the mass spectrometer, thus the samples can be combined and analysed together, *e.g. via* LC-MS.

In this chapter, we focus on one approach that has become widely used among the scientific community over the last years. It makes use of the metabolic incorporation of different isotopically labelled forms of amino acids and is called stable isotope labelling with amino acids in cell culture (SILAC).[8] The SILAC method is a very simple and easy-to-use approach as the different stable isotopic forms of amino acids are metabolically incorporated into the proteins by using special cell culture media. This has the additional advantage that the samples to be compared can be mixed directly before cell lysis, circumventing quantitative errors due to sample losses during the sample preparation processes. The relative quantitative information of peptides can be directly deduced from the MS spectrum by comparing the signal intensities of the coeluting isotopic peptide partners. The identification of the corresponding peptide is achieved by tandem MS (MS/MS).[8,9] As indicated by its name, SILAC was initially developed for cell culture experiments, clearly restricting its applications. Recent developments now allow the labelling of whole model organisms such as bacteria,[10] yeast,[11] flies,[12] and even mice.[13] Moreover, a very elegant way to quantitatively compare samples such as cancer tissue was recently described.[14] The so-called 'Super-SILAC' approach makes use of a repertoire of SILAC-labelled cell cultures which serve as an internal standard that is mixed with the tissue samples, making the SILAC method even more widely applicable.

8.2 Materials

8.2.1 SILAC Labelling

- Cell line of choice, adherent (*e.g.* HeLa cells) or suspension cells (*e.g.* Jurkat T cells)
- Essential amino acids containing different stable isotopes of carbon, nitrogen or hydrogen (isotopologues), *e.g.* L-arginine, L-lysine, L-methionine, L-leucine; from *e.g.* Sigma, Silantes, Cambridge Isotopes, Eurisotop (Note 1)
- Medium devoid of amino acids that are used as isotopologues, *e.g.* DMEM, RPMI for SILAC from *e.g.* GIBCO, PAA (Note 2)
- Filtering device for medium preparation; *e.g.* 0.22 µm PES bottle-top vacuum filters (Corning)
- Sterile dialyzed fetal bovine serum (FBS), filtered against a 10 kDa cut-off (Invitrogen)
- Antibiotics: penicillin, streptomycin (100×, Invitrogen); others if necessary
- L-Glutamine (100×, Invitrogen); if not already contained within the medium
- Trypsin for detaching adherent cells (*e.g.* Trypsin-EDTA, GIBCO or PAA)

8.2.2 Sample Preparation

- Denaturing lysis buffer containing *e.g.* 6 M urea, 2 M thiourea[15] or 4% SDS[16]
- Dithiothreitol (DTT) and iodoacetamide (IAA) for protein reduction and alkylation
- Protease for proteolytic cleavage of proteins (*e.g.* sequencing grade modified trypsin, Promega) depending on the labelled amino acids
- Materials and buffers depending on sample separation procedures

8.2.3 LC-MS Analysis

- Reversed phase (RP) C-18 material for desalting the peptide mixture, *e.g.* by using StageTips[17]
- Trifluoroacetic acid (TFA), acetic acid (AcOH) and acetonitrile (ACN) for desalting and chromatography buffers
- Silica emitters packed with reversed-phase C-18 material, *e.g.* 3 μm Reprosil C-18 beads (Dr. Maisch); 15 cm 75 μm ID as chromatography column (Proxeon, NewObjective)
- Hydrophilic solvent, *e.g.* 0.5% AcOH
- Hydrophobic solvent, *e.g.* 80% ACN, 0.5% AcOH

8.2.4 Equipment

- Nano-HPLC (*e.g.* Proxeon, Agilent, Waters, Eksigent, Dionex) on-line coupled to the mass spectrometer
- Mass spectrometer, *e.g.* LTQ Orbitrap Velos (Thermo Fisher Scientific) or QSTAR Elite (AB SCIEX)
- MS analysis software (*e.g.* MaxQuant,[18] Census,[19] MassLynx (Waters), ProteinProphet (ISB, Seattle), MS Quant,[20] Proteome Discoverer (Thermo Fisher Scientific)

8.3 Methods

SILAC relies on the metabolic incorporation of isotopically labelled amino acids into the proteome. In most SILAC experiments arginine (Arg) and lysine (Lys) are chosen, as trypsin is generally the protease of choice to proteolytically cleave proteins into peptides prior to mass spectrometric analysis. This guarantees that almost all generated peptides contain isotopic counterparts because trypsin cleaves C-terminally to Arg and Lys. Depending on the study design two (double labelling) to a maximum of three (triple labelling) different conditions are normally compared in one experiment. For both amino acids two different isotopic forms, one medium (Arg6, LysD4) and one heavy labelled (Arg10, Lys8), are commercially available besides the natural light analogues (Arg0, Lys0). If more conditions are thought to be compared, *e.g.* as in time scale experiments, several SILAC experiments can be combined *via* one common condition shared between all triplicates.[21]

8.3.1 Preparation of SILAC Medium

For the SILAC labelling process, media have to be generated that contain the different isotopic forms of the amino acids of choice (Table 8.1). Prepare the different SILAC media (light, medium and heavy) exactly the same way to avoid side effects caused by different culture conditions.

- Calculate carefully how much medium has to be prepared
  - ⓘ Standard media such as DMEM or RPMI devoid of the isotopically labelled amino acids used in the experiment (*e.g.* Lys and Arg) can be purchased from different companies. Also, unconventional or customer-designed media can be purchased if bigger amounts are ordered
  - ⓘ It is advisable to prepare highly concentrated stock solutions of amino acids to avoid dilution of the medium
- Prepare stock solutions of *e.g.* Arg and Lys for both light and heavy amino acids in PBS or non-restituted culture medium (*e.g.* 84 mg ml^{-1} for Arg and 146 mg ml^{-1} for Lys) (Note 3). Moreover, it is recommended to buy SILAC amino acids in larger amounts to avoid batch to batch variations
- Mix medium, amino acids and necessary ingredients (*e.g.* antibiotics, L-glutamine or pyruvate) in the desired concentrations and filter the prepared medium with a sterile filter (*e.g.* 0.22 μm PES filter) (Note 4)
  - ⓘ If the medium is already sterile it is sufficient to filter only non-sterile components, such as amino acids, antibiotics and glutamine, resuspended in a small volume of medium. Afterwards, add the filtered ingredients to the remaining medium
- The final concentrations of Arg and Lys have to be adjusted for each cell line to achieve optimal labelling efficiencies and to suppress Arg to proline (Pro) conversion (see 8.4.1). As starting conditions dilute your amino acid stock solutions *e.g.* 1:3000 (for Arg) and 1:2000 (for Lys) resulting in concentrations of 28 mg L^{-1} and 73 mg L^{-1}, respectively
- Add sterile dialyzed FBS in a concentration of *e.g.* 10% to the filtered medium

Table 8.1 Composition of standard SILAC medium (V = 500 mL)

Reagent	Volume (dilution factor)	Final concentration
RPMI/DMEM SILAC medium	439.6 ml	—
Pen/Strep (100×)	5 ml	1×
L-Glutamine (100×)	5 ml	1×
L-Arginine (84 mg/ml)	166.6 μl (1:3000)	28 mg/mL
L-Lysine (146 mg/ml)	250.0 μl (1:2000)	73 mg/mL
Dialysed FBS (10×), after filtering	50 ml	10%

ⓘ Do not add FBS before filtering to avoid clogging of the filter (Note 5)

ⓘ Never use non-dialyzed serum as this contains amino acids such as Arg and Lys which would contaminate your SILAC medium with the light isotopologues, hampering sufficient label incorporation (Note 6)

■ Store the medium at 4 °C up to 2 months until further use.

8.3.2 SILAC Labelling and Incorporation Test

One big advantage of SILAC is its simplicity and accuracy. In comparison to conventional non-quantitative experiments, only SILAC medium has to be prepared additionally as described above.

- Start growing your cells in conventional medium. Once cells are adapted to the growth conditions, the medium can be replaced by the different SILAC media. After washing once with PBS, detach adherent cells from cell culture plates, *e.g.* by adding trypsin. Stop trypsin activity immediately when cells have detached from the culturing plates by adding fresh medium. Centrifuge the suspension (5 min, 400 *g*, RT[15]) and resuspend the cells in fresh medium. Suspension cells are split by centrifugation and subsequent resuspension in an appropriate volume of fresh SILAC medium

  ⓘ It is important to carefully remove non-SILAC medium and trypsin solution as they contain natural, non-labelled forms of amino acids and would affect optimal label incorporation

  ⓘ It is advisable to check before, if the cells grow in medium supplemented with dialysed FBS (see Notes 5 and 6). In general, SILAC labelling should not affect cell morphology or growth rates if properly applied.[15] If the cells show different behaviours when switched to SILAC media, this can be taken as an indicator for wrong labelling conditions

- Culture cells in SILAC medium for about 5–10 doublings, split the cells as recommended (*e.g.* every 2–3 days with splitting ratios of about 1:4). Start labelling the cells in smaller subcultures to save medium

  ⓘ Following these instructions a minimal label incorporation of 95% should be achieved

  ⓘ Labelled cells can be frozen and thawed again when needed

  ⓘ An amount of about 1×10^7 cells concentrated in 1 mL freezing solution (*e.g.* 10% DSMO in dialysed FBS) is recommended for many cell lines. Freezing labelled cells can speed up the whole labelling process and saves money as the initial labelling steps are avoided

  ⓘ Before starting the SILAC experiment it is advisable to first check the incorporation efficiency of the labelled amino acids into the proteins. For this reason, use a minimum of about 1×10^6 cells for standard incorporation testing

  ⓘ Only medium or heavily labelled cells have to be checked for labelling efficiencies, since the conversion is calculated *via* comparing the heavy to the light form of the isotopic label. The determined concentrations are then adapted to the light condition

- ⓘ Incomplete incorporation (labelling efficiencies below 95%) has to be avoided to gain the best accuracy of quantification. Moreover, it is much more convenient to adjust incomplete labelling before than correcting the data afterwards
- ⓘ Depending on the cell type and medium conditions the supplemented heavy Arg might be metabolically converted into Pro, giving rise to additional forms of isotopic peptides. As heavy Pro can contribute to up to 30–40% of Pro-containing peptide abundances,[22] it is important to check also the conversion of Arg to Pro besides the general incorporation
- Harvest the cells and wash carefully with ice-cold PBS to remove serum proteins originating from the medium. Keep the cells on ice during the whole process. Store cell pellets in liquid nitrogen or go on directly with the sample preparation
- Lyse the cells *e.g.* in urea or detergent-containing buffers. Digest the extracted proteins with trypsin and subject the generated proteolytic peptides to LC-MS analysis using standard mass spectrometric protocols[15] (and see also Chapters 5 and 6)
  - ⓘ For labelling checks, no extensive separation of the protein or peptide sample is needed. In fact, it is sufficient to analyse several hundred isotopically labelled peptides to calculate the incorporation efficiency
  - ⓘ Sample amounts of about 1 µg per LC-MS run are sufficient using a 60-min gradient on the described LC-MS platform
- If satisfactory label incorporation was achieved (see section 8.4.1), grow the culture to the appropriate cell number and use them for the SILAC experiment

8.3.3 SILAC Experiment

In the following section we will explain the proceedings for a SILAC double labelling experiment; however, this can be easily extended to triple labelling. Once cells are grown in SILAC medium and labelling checks reveal sufficient labelling incorporation the experiment can be started. Avoid any stress conditions for the cells, since this could lead to side effects in protein regulations.

- Treat an appropriate amount of cells either with mock or specific stimuli, like growth factors, inhibitors, *etc.*
  - ⓘ For a normal proteome experiment 1 mg of protein is usually sufficient. One 15 cm dish of HeLa cells (grown to 70–80% confluency) should contain about 2–3 mg of protein. Be aware that the amount of extracted protein varies substantially from cell line to cell line
  - ⓘ Several replicates of the same experiment are recommended where labels are swapped across the conditions. This gives additional information about the labelling and possible contaminants (Note 7). Moreover, biological and technical replicates should be performed for sound statistics and increased protein identifications. In general at

least three replicates should be performed to allow the application of statistical tools
- Stop the treatment by putting cells on ice. Remove the medium and wash the cells two or three times with ice-cold PBS. Depending on the experiment merge the two populations directly on cell or protein level. Extract the proteins by lysing the cells with detergent or urea containing buffers (Note 8)
  - (i) If the samples should be combined on the cell level, cell numbers have to be equalized (*e.g.* by cell counting) before treatment
  - (i) When combining the samples after cell lysis, protein concentrations have to be determined and equal protein amounts are mixed (Note 9)
  - (i) Adherent cells may be lysed directly on the culture plates by adding denaturing buffers after washing with PBS. This is the fastest way to lyse the cells after treatment
  - (i) Suspension cells can be easily centrifuged at 4 °C. After the first PBS washing step, cells from the different conditions may be combined if same cell numbers are used
  - (i) Cell pellets can be frozen and stored in liquid nitrogen
  - (i) Use separation techniques to reduce the complexity of the protein samples especially when working with whole cell lysates, containing thousands of different proteins and protein species
  - (i) Several methods can be used either on the protein level (*e.g.* SDS-PAGE[23]) or peptide level (*e.g.* Offgel,[24] ion exchange chromatography[25]) after proteolytic digestion
- Digest the proteins with an appropriate protease
  - (i) As stated before, the proteolytic digestion of proteins is commonly carried out with trypsin. This protease bears several advantages as it generates on average well-suited peptide lengths for the mass spectrometric analysis. Additionally, the peptides carry C-terminally a positively charged amino acid (Arg or Lys) supporting the ionization of the peptides
- Desalt the peptide solution before mass spectrometric analysis
  - (i) As the digestion step is carried out in buffers containing detergents and salts it is absolutely necessary to purify the peptide sample before LC-MS analysis. In case no online precolumns are used within the LC-MS setup, micropurification strategies to handle small sample amounts tailored for samples designated to LC-MS analysis have to be applied[17]
- Bring the desalted peptides into a reversed phase compatible milieu by resuspending the sample in acidified water (*e.g.* 0.1% TFA or 0.5% AcOH), containing maximal concentrations of organic solvent of 2–5% (*e.g.* ACN)
- Load per LC-MS analysis peptide amounts of about 2–5 µg on the reversed-phase column, which is coupled online to the mass spectrometer
  - (i) It is recommended to split the sample into at least two parts if enough material is available. This offers the opportunity to reanalyze sample replicates

- Define the chromatographic gradient depending on the complexity and composition of the samples to be analysed
  - ⓘ In general, gradient lengths of 90–120 min (*e.g.* from 2 to 30% hydrophobic solvent) are recommended
  - ⓘ Most mass spectrometers used today for quantitative proteomics, such as hybrid MS instruments (*e.g.* LTQ-Orbitrap, Q-TOF), offer different analysis methods varying from high resolution and high mass accuracy to low resolution scan types. Additionally, variable fragmentation modes for peptide dissociation are available.[26]
  - ⓘ Acquisition of high-resolution MS spectra is necessary in the first place for precise peptide quantification, as coeluting nearly isobaric peptides can still be distinguished and thereby correctly quantified in the MS spectrum. In addition, high mass accuracy is beneficial for the peptide identification process
  - ⓘ For the acquisition of MS/MS spectra different fragmentation types are available. Commonly, peptide fragmentation is achieved by collision activated dissociation (CID, HCD). Recently, new fragmentation modes have been developed relying on chemical and thereby 'milder' dissociation processes preferentially used for the analysis of post-translational modifications (ECD, ETD)
  - ⓘ Try to find a good balance between best identification and quantification results. In general, the more MS/MS spectra are acquired the more peptides can be identified. However, the accuracy in quantification might be negatively affected since less MS spectra are recorded which deliver the quantitative information (Note 10)

8.4 Data Analysis

For analysing SILAC LC-MS/MS data several software packages have been developed. In this section, we will not go into details of specific software platforms,[18–20] but give general recommendations for the analysis of SILAC data.

8.4.1 Labelling Check

Complete incorporation of the isotopic amino acids into 'heavy cell cultures' is, as mentioned above, important for accurate quantification by SILAC. Also, the occurrence of Arg to Pro conversion has to be evaluated.

- First, manually inspect the MS output files to check labelling efficiencies. Select several peaks along the chromatogram and check whether the unlabelled light peptide almost disappeared (see Figure 8.1)
- Analyse your MS data with a suitable software package to identify the sequenced peptides and to calculate the peptide ratios for all analysed isotopic peptide pairs (as in a normal double labelling experiment—see section 8.4.2). Extract all calculated SILAC ratios of the quantified peptides from the output files

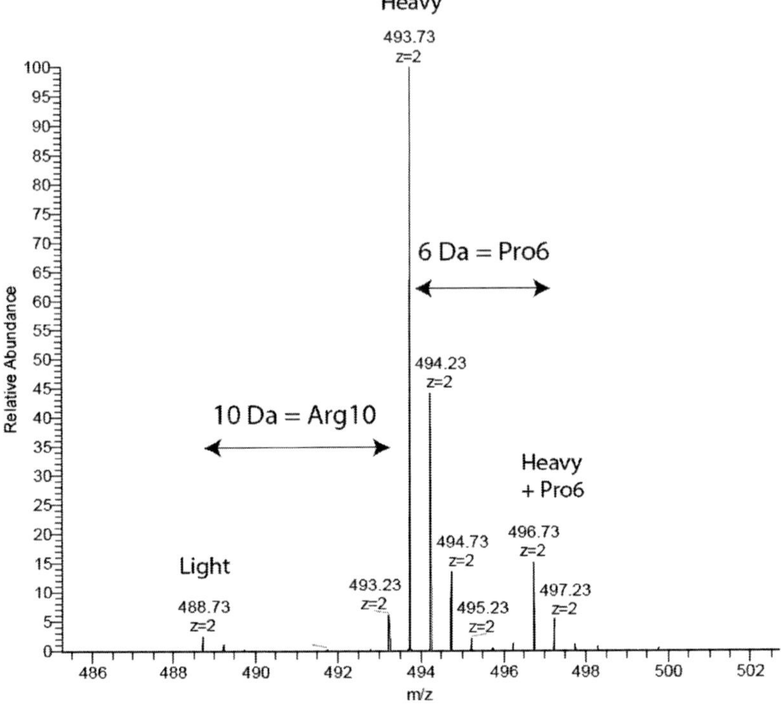

Figure 8.1   Example spectrum of a SILAC labelled peptide signal. The most intense peak corresponds to the heavy labelled peptide. Towards the lower mass region the unlabelled peptide is still visible with very low intensity indicating good incorporation efficiency. However, an additional signal with a mass difference of 6 Da to the heavy labelled peptide appears in the higher mass region denoting Arg to Pro conversion.

ⓘ Some software packages allow ratio normalization steps over all SILAC peptide pairs to compensate *e.g.* for mixing errors after combining light and heavy samples. The normalization acts on the assumption that most SILAC ratios are around 1, which is not the case when only analysing heavy samples for incorporation tests. It is therefore important to use non-normalized ratios for calculating labelling efficiencies

- Calculate the average over all peptide ratios and define the incorporation rate by using the following formula:

$$\text{average incorporation } (\%) = [1 - (1/\text{mean} + 1)] \times 100 \qquad (8.1)$$

ⓘ The incorporation efficiency should reach at least 95%

- For a more sophisticated analysis of labelling efficiencies, split Arg- and Lys-containing peptides to get separate incorporation efficiencies for these two populations
- Exclude missed cleavages as a search parameter in order to avoid peptides containing both Arg and Lys

- In the following, an R script (http://www.r-project.org/) is provided which can be used to analyse the incorporation rate of SILAC labelled samples (Figure 8.2). It can be applied to tab-delimited .txt-files containing the columns 'Ratio H/L' (peptide ratios between H and L), 'R Count' (sum of Arg residues included in the corresponding peptide) and 'K Count' (sum of Lys residues included in the corresponding peptide)
- Interpret the incorporation rates of Arg and Lys on the basis of the density plots delivering the average labelling efficiency and the broadness of both populations (see Figure 8.3)

```r
setwd(choose.dir())

peptides <- read.table("FILENAME.txt", quote = "\"", header = TRUE,

sep = "\t", stringsAsFactors = FALSE, comment.char = "")

filteredpeptides <- peptides[is.na(peptides$Ratio.H.L) != TRUE,]

Rpeptides <- filteredpeptides[filteredpeptides$R.Count > 0,]

Rpeptides <- Rpeptides[Rpeptides$K.Count == 0,]

Kpeptides <- filteredpeptides[filteredpeptides$K.Count > 0,]

Kpeptides <- Kpeptides[Kpeptides$R.Count == 0,]

plot( density(1- 1/(filteredpeptides$Ratio.H.L+1)),

     xlim=c(0.8,1.0),

     col="black",

     ylab=expression("Density"),

ylim=c(0,30), main="Incorporation rate")

lines(density(1-1/(Kpeptides$Ratio.H.L + 1)), col="green")

lines(density(1-1/(Rpeptides$Ratio.H.L + 1)), col="red")

den <- density(1- 1/(filteredpeptides$Ratio.H.L+1))

inc <- den$x[which.max(den$y)]

denK <- density(1- 1/(Kpeptides$Ratio.H.L+1))

incK <- denK$x[which.max(denK$y)]

denR <- density(1- 1/(Rpeptides$Ratio.H.L+1))

incR <- denR$x[which.max(denR$y)]

pep <- paste("All peptides: ", round(inc,3))

arg <- paste("Arginine peptides: ", round(incR,3))

lys <- paste("Lysine peptides: ", round(incK,3))

legend(x="topleft", c(pep, arg, lys), col=c("black", "red", "green"),lwd=1)
```

Figure 8.2 R script to calculate separate incorporation efficiency of Arg- and Lys-containing peptides.

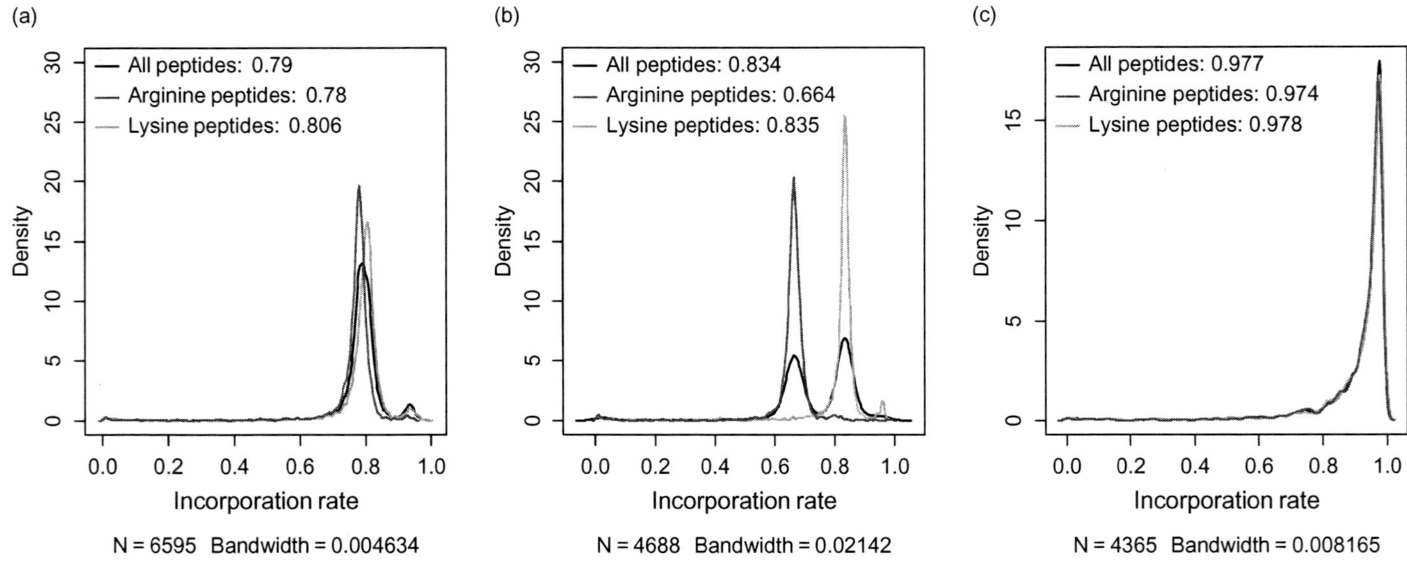

Figure 8.3 Density plots of different incorporation efficiencies for Arg and Lys containing peptides: (a) Incomplete incorporation for Arg and Lys both showing similar labelling efficiencies. (b) Incomplete labelling of Arg and Lys both showing different incorporation behaviours (split populations). (c) Complete labelling for both Arg and Lys.

- In case Arg or Lys is not fully incorporated, the concentration in the medium has to be adjusted and labelling checks have to be repeated. In the case of incomplete labelling, light Pro would convert to light Arg. Conversely, Arg could have been fully incorporated but then converted to Pro, since Arg is added in excess (see below)
  - ⓘ Be aware when Arg is used as heavy amino acid that it can be metabolically converted to heavy Pro, giving rise to additional peptide species
- First, manually check Pro-containing peptide peaks within the chromatogram. Arg6 converts to Pro5, Arg10 to Pro6, leading to peptides containing both heavy Arg and heavy Pro. Consequently, for doubly charged peptides ($z = 2$), an additional peak 3 or 2.5 Thomson (Th) away from the heavy, non-converted peptide should appear (Figure 8.1)
  - ⓘ Besides manual inspection of the MS files, the Arg to Pro conversion can also be monitored by the data analysis software
- Define heavy Pro (Pro6 and Pro5 for heavy and medium labels, respectively) as a variable modification for the database search. The amount of sequenced heavy Pro-containing peptides can serve as an estimate of the extent to which the conversion had occurred

If Arg to Pro conversion is observed, different strategies can be applied:

- Reduce the Arg concentration (or increase the Pro concentration—less recommended) until the conversion to Pro becomes unfavourable. Try to find the correct balance between Arg and Pro concentrations to achieve optimal incorporation efficiencies and not to negatively affect growth rates of the cells.[27]
- Correct the ratios by adding the contribution from heavy Pro to the heavy Arg peak[22] (implemented in Census software)
- Substitute normal Arg by $^{15}N_4$-Arg in the light condition. Using this strategy, the Pro conversion also occurs in the light condition and therefore the signals of the unconverted peptides can be directly compared without introducing an additional quantitative error.[28]
  - ⓘ Different cell lines may show different incorporation rates and Arg conversion behaviours. It might happen that some cells do not allow complete label incorporation without Arg to Pro conversion, so that a compromise between complete incorporation and a certain extent of converted Arg has to be tolerated

8.4.2 SILAC Experiment

Besides the correct identification of the analysed peptides, in SILAC experiments the accurate ratio determination of the isotopic peptide pairs has to be ensured over large data sets. In the following, we highlight several aspects within the data analysis workflow which are important to achieve optimal results.

8.4.2.1 Software Analysis

- Define appropriate search parameters in the data analysis software
- In general, it is important to specify the protease used, fixed and variable modifications, missed cleavages, as well as thresholds for mass tolerances

and FDR calculations. Moreover, it is possible to define how many peptides have to be identified for reliable protein identifications or protein quantifications
- Use a database search approach that allows the reliable estimation of false positive rates and the occurrence of common contaminations in your dataset
    - ⓘ An easy way to estimate false-positive identifications is the use of concatenated *forward* and *reverse* databases. These databases contain, in addition to the normal proteins, all proteins with reversed sequence from the N- to the C-terminus. By searching against such databases the amount of false-positive identifications can be easily estimated based on the number of identifications from normal and reverse entries.[29]
    - ⓘ Moreover, known contaminants like keratins or the protease used, should be included and specifically flagged in the database. This facilitates the subsequent filtering steps during data analysis
- Define the correct labels used in the experiment
- The analysis can be directed by sequenced and identified peptides. For those peptides identified, the isotopic ratios are calculated and used as the basis for protein quantification. Here the heavy SILAC amino acids used have to be selected as variable modifications
- A second approach makes use of the different but characteristic mass gaps between SILAC peptide pairs (*e.g.* 6 and 10 Da for peptides containing heavy Arg6/Arg10 or 4 and 8 Da for peptides carrying heavy Lys4/Lys8). Here the detection of the isotopic peptides takes place even before the identification step, and this has several advantages: (1) the different isotopic peptide forms (*e.g.* light, medium, and heavy labelled peptides) are grouped and consequently the corresponding isotopic amino acids can be selected as fixed modification subsets; (2) the characteristic mass gaps support peptide identification by distinguishing between Arg- and Lys-containing peptides

In principle, SILAC peptide pairs have the same physicochemical properties and consequently coelute from the chromatographic column into the mass spectrometer, when using ^{13}C and ^{15}N as isotopic markers. However, if the amino acid is substituted with deuterium (*e.g.* Lys4 contains 4 deuterium atoms), the deuterated peptides elute slightly earlier when using reversed-phase chromatography due to their slightly more hydrophilic character compared to their non-deuterated counterparts.

- ⓘ Peptide quantification based only on single MS spectra might lead to severe errors for deuterated SILAC peptide pairs
- ⓘ It is recommended to use the so-called extracted ion chromatogram (EIC) as three-dimensional peak profiles for quantification. Here, slight retention time variations do not negatively affect quantification. In addition, the calculation of the peptide ratios is much more precise, also for non-deuterated peptides, as the calculation is integrated over the complete elution profile

8.4.2.2   Evaluation of the Output Data

- First check the technical performance of your LC-MS/MS setup and the results of your SILAC quantification analysis
  - ⓘ Depending on the source, complexity, the amount of analysed sample, the LC-MS hardware and applied bioinformatic software package, approximately 20–60% of all acquired MS/MS spectra can lead to peptide identifications. Lower rates might indicate reduced performance of the machine or inappropriate settings for the search
- Remove entries corresponding to contaminants and reverse entries from your output tables (see section 8.4.2.1)
- After this filtering step, SILAC pairs can be sorted according to their H/L ratios. Keep in mind that on a linear scale down-regulations have ratios between 0 and 1 and up-regulations between 1 and infinity. It is therefore advisable to transform ratios into logarithmic values to achieve an unbiased illustration of up- and down-regulations
- Using box plots, density plots or histograms (see Figure 8.4A–C) can give a first global overview of the extent to which quantitative protein changes has occurred in the analysed samples. Furthermore, you can check for normal distribution of the population and get an overview of the number of regulated events. More sophisticated information about the population itself and the amount of strongly regulated proteins can be obtained *e.g.* by plotting the ratios against the summed intensities (Figure 8.4D)
- In addition, significances of the determined protein ratios should be calculated. Assuming a normal distribution of logarithmic ratios, left and right standard deviations of the whole population are evaluated. Based on these parameters, significances of single ratios can be calculated. Hereby, uniform values for regulations are given, which can be combined with any other population (see Figure 8.4D)
  - ⓘ For significant regulations it is recommended to include the intensity of the proteins in the ratio calculation. Since more intense regulations are detected with better accuracy, an intensity-weighted algorithm can regulate for these differences over the intensity scale[18]
  - ⓘ In general, popular levels of significances are 5, 1 and 0.1%
- To get information about which cellular phenomena are affected by the significant regulations, analyse the regulated proteins with regard to their involvement in specific cellular processes, pathways or networks by using tools such as Gene Ontology (http://www.geneontology.org), KEGG (http://www.genome.jp/kegg/) or STRING (http://string.embl.de/)
- Moreover, further information can be extracted by enrichment testing or hierarchical clustering of several experiments
- Regulations over several experiments can be further analysed, *e.g.* by t-tests or analyses of variance (ANOVA) (in case of more than two groups)
- Reproducibility of biological or technical replicates should be examined, to emphasize the regulations derived from the treatment in comparison to technical variations

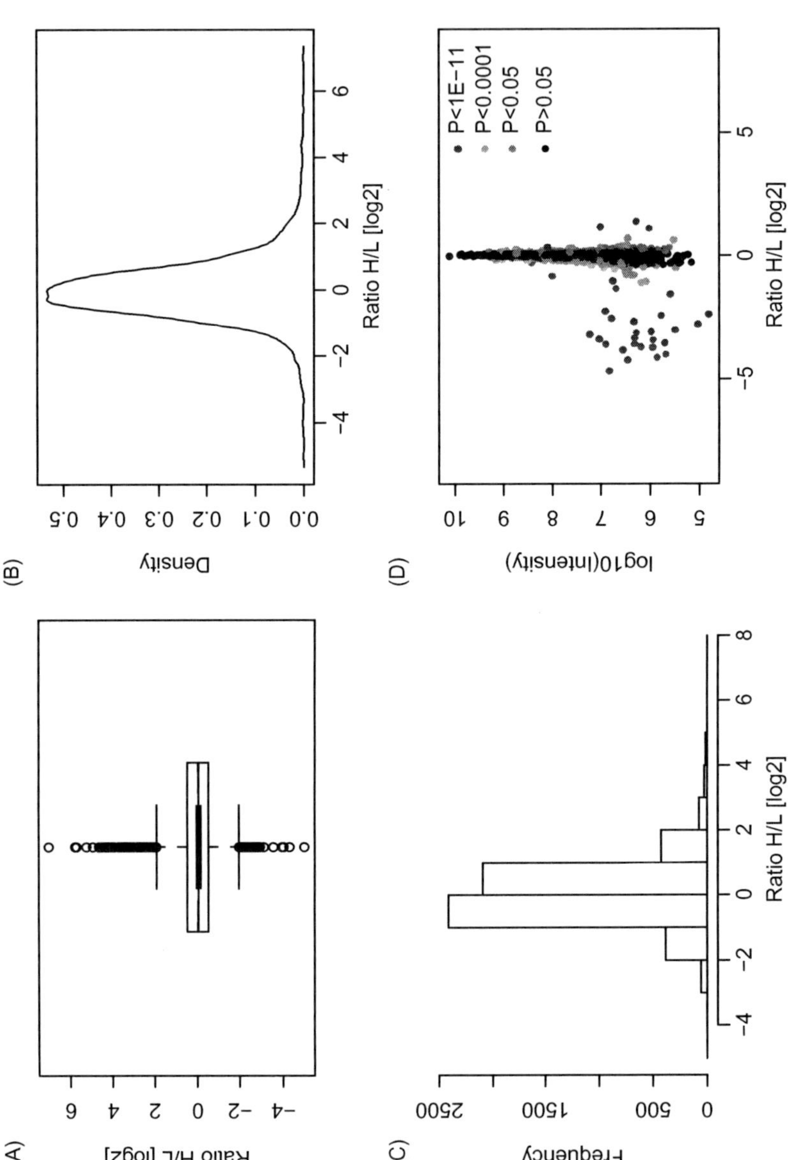

Figure 8.4   Different graphical presentations to depict global quantitative protein changes within the analysed dataset. These plots may help to give a first impression of the quality of the SILAC experiment.

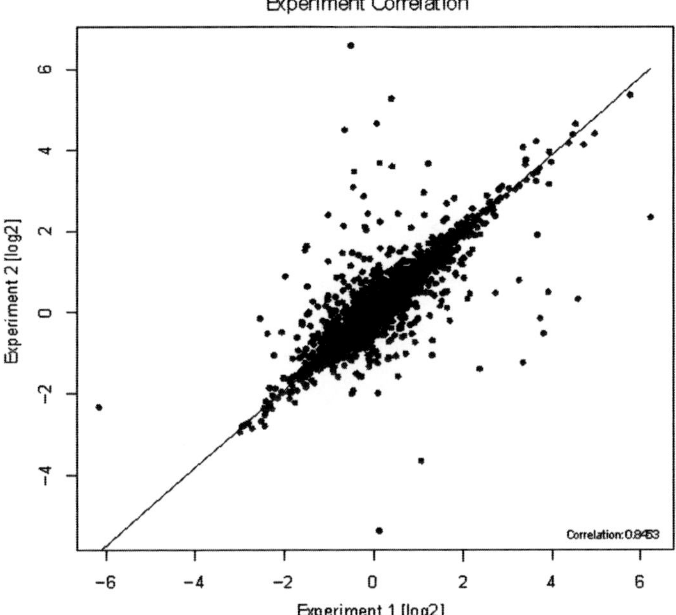

Figure 8.5 Correlation between two different technical replicates of the same experiment. The given correlation corresponds to the Pearson correlation coefficient.

- Correlations between experiments can be calculated by plotting the ratios against each other using a scatter plot. The Pearson correlation coefficient can be calculated giving a measure of the linear dependence between two variables (Figure 8.5)

8.5 Notes

1. In principle, only essential amino acids, which cannot be metabolically synthesized by the cultured cells, should be used for labeling. By adding the amino acids to the medium these should serve as the only source for the particular amino acid.
2. No component of the medium may contain any additional amino acid source. Additional sources of amino acids (*e.g.* in serum) have to be carefully accounted for, to avoid incomplete labeling.
3. Take care of the water/HCl (1× HCl, 2× HCl) content in the amino acid powder, which may vary between different batches. Moreover, amino acids from different batches or vendors might have different isotopic purities.
4. Media usually contain a large quantity of amino acids and other nutrients. However, to reduce costs and keep the balance of

non-essential amino acids, the concentration of the added amino acids is scaled down to the necessary level.

5. FBS could also be excluded from the medium stock and added during each splitting step in the proper amount to prolong the stability of the medium to several months.

6. Add purified growth factors or a small amount of normal serum, if cells do not grow in media supplemented with dialysed serum. Furthermore, dialysed serum generated with a less stringent mass cut-off filter may be tested.

7. During all steps of the experiments it is highly recommended to avoid any contamination of samples by keratins, *etc.* since these interfere with protein identification by avoiding sequencing of the real sample peptides. However, contaminants also have an advantage, since they can only be detected in the light form and never in the heavy state. Like this, under 'normal labelling' conditions they show a very low ratio for H/L (see Figure 8.4D). If a reverse experiment is being performed, the contaminant entries should still occur in the low ratio area. In contrast, all other changes which are related to the treatment should swap together with the label.

8. Depending on the experiment, it might be highly advisable to add protease, phosphatase or kinase inhibitors to the lysis buffer to avoid side effects caused by enzymes that are not inactivated.

9. Approaches for measuring protein concentrations have to be selected according to the sample buffer. For many buffer systems, Bradford or Lowry methods are sufficient; however, when SDS is used as an additive for protein solubilization, alternative methods have to be chosen, *e.g.* tryptophan-fluorescence measurements.

10. If LTQ-Orbitrap instruments are used, low-resolution ion-trap CID MS/MS spectra can be acquired in parallel to high-resolution survey MS scans in the Orbitrap mass analyser. In principle, for low-resolution CID measurements Top5 to Top10 methods with LTQ-Orbitrap XL mass spectrometers and Top10 to Top15 methods for LTQ-Orbitrap-Velos machines are recommended.[30] If the fragmentation is carried out in the HCD collision cell,[26] fragmentation spectra have to be recorded in the Orbitrap mass analyser after the preceding high-resolution survey MS scan. Consequently, fewer MS/MS scans can be recorded in one cycle.

References

1. R. Aebersold and M. Mann, *Nature*, 2003, **422**, 198.
2. S. E. Ong and M. Mann, *Nat. Chem. Biol.*, 2005, **1**, 252.
3. M. H. Elliott, D. S. Smith, C. E. Parker and C. Borchers, *J. Mass. Spectrom.*, 2009, **44**, 1637.

4. W. X. Schulze and B. Usadel, *Annu. Rev. Plant Biol.*, **61**, 491.
5. S. P. Gygi, B. Rist, S. A. Gerber, F. Turecek, M. H. Gelb and R. Aebersold, *Nat. Biotechnol.*, 1999, **17**, 994.
6. X. Yao, A. Freas, J. Ramirez, P. A. Demirev and C. Fenselau, *Anal. Chem.*, 2001, **73**, 2836.
7. Y. Oda, K. Huang, F. R. Cross, D. Cowburn and B. T. Chait, *Proc. Natl. Acad. Sci. U S A.*, 1999, **96**, 6591.
8. S. E. Ong, B. Blagoev, I. Kratchmarova, D. B. Kristensen, H. Steen, A. Pandey and M. Mann, *Mol. Cell Proteomics*, 2002, **1**, 376.
9. H. Steen and M. Mann, *Nat. Rev. Mol. Cell Biol.*, 2004, **5**, 699.
10. B. Soufi, C. Kumar, F. Gnad, M. Mann, I. Mijakovic and B. Macek, *J. Proteome Res.*, **9**, 3638.
11. A. Gruhler, J. V. Olsen, S. Mohammed, P. Mortensen, N. J. Faergeman, M. Mann and O. N. Jensen, *Mol. Cell. Proteomics*, 2005, **4**, 310.
12. M. D. Sury, J. X. Chen and M. Selbach, *Mol. Cell Proteomics*, 2010, **9**, 2173.
13. M. Kruger, M. Moser, S. Ussar, I. Thievessen, C. A. Luber, F. Forner, S. Schmidt, S. Zanivan, R. Fassler and M. Mann, *Cell*, 2008, **134**, 353.
14. T. Geiger, J. Cox, P. Ostasiewicz, J. R. Wisniewski and M. Mann, *Nat. Methods*, **7**, 383.
15. S. E. Ong and M. Mann, *Nat. Protoc.*, 2006, **1**, 2650.
16. J. R. Wisniewski, A. Zougman, N. Nagaraj and M. Mann, *Nat. Methods*, 2009, **6**, 359.
17. J. Rappsilber, M. Mann and Y. Ishihama, *Nat. Protoc.*, 2007, **2**, 1896.
18. J. Cox and M. Mann, *Nat. Biotechnol.*, 2008, **26**, 1367.
19. S. K. Park, J. D. Venable, T. Xu and J. R. Yates, 3rd, *Nat. Methods*, 2008, **5**, 319.
20. P. Mortensen, J. W. Gouw, J. V. Olsen, S. E. Ong, K. T. Rigbolt, J. Bunkenborg, J. Cox, L. J. Foster, A. J. Heck, B. Blagoev, J. S. Andersen and M. Mann, *J. Proteome Res.*, **9**, 393.
21. J. V. Olsen, B. Blagoev, F. Gnad, B. Macek, C. Kumar, P. Mortensen and M. Mann, *Cell*, 2006, **127**, 635.
22. S. K. Park, L. Liao, J. Y. Kim and J. R. Yates, 3rd, *Nat. Methods*, 2009, **6**, 184.
23. A. Shevchenko, H. Tomas, J. Havlis, J. V. Olsen and M. Mann, *Nat. Protoc.*, 2006, **1**, 2856.
24. N. C. Hubner, S. Ren and M. Mann, *Proteomics*, 2008, **8**, 4862.
25. J. R. Wisniewski, A. Zougman and M. Mann, *J. Proteome Res.*, 2009, **8**, 5674.
26. J. V. Olsen, B. Macek, O. Lange, A. Makarov, S. Horning and M. Mann, *Nat. Methods*, 2007, **4**, 709.
27. T. A. Prokhorova, K. T. Rigbolt, P. T. Johansen, J. Henningsen, I. Kratchmarova, M. Kassem and B. Blagoev, *Mol. Cell. Proteomics*, 2009, **8**, 959.

28. D. Van Hoof, M. W. Pinkse, D. W. Oostwaard, C. L. Mummery, A. J. Heck and J. Krijgsveld, *Nat. Methods*, 2007, **4**, 677.
29. J. E. Elias and S. P. Gygi, *Nat. Methods*, 2007, **4**, 207.
30. J. V. Olsen, J. C. Schwartz, J. Griep-Raming, M. L. Nielsen, E. Damoc, E. Denisov, O. Lange, P. Remes, D. Taylor, M. Splendore, E. R. Wouters, M. Senko, A. Makarov, M. Mann and S. Horning, *Mol. Cell. Proteomics*, 2009, **8**, 2759.

LC-MS for the Identification of Post-Translational Modifications of Proteins

BORIS MACEK

Proteome Center Tübingen, Interdepartmental Institute for Cell Biology, University of Tübingen, Auf der Morgenstelle 15, 72076 Tübingen, Germany

9.1 Introduction

Protein function is determined by its linear amino acid sequence, additional levels of structural organization, and post-translational modifications (PTMs) that are covalently linked to the side chains of amino acid residues. PTMs are present in every kingdom of life and almost every protein is likely to be modified in some way during its lifetime. Up to 2000 genes (10%) in the human genome are involved in PTM biology and more than 200 PTMs have been reported to date.[1] In the postgenomic era, PTMs are gaining attention as they can profoundly influence structure and function of gene products (proteins) and their biology can only partly be studied at the genome level.

PTMs are characterized by their chemical and physical diversity. Some are rather simple (deamidation, methylation, acetylation, phosphorylation), others are complex (*N*-glycosylation, GPI-anchors). Depending on their composition they can be hydrophilic (glycosylation), hydrophobic (myristoylation), or charged (phosphorylation, acetylation, some forms of glycosylation). This chemical diversity presents a range of issues in PTM analysis, especially in biochemical enrichment and separation of modified proteins or peptides. PTMs also differ in their

RSC Chromatography Monographs No. 15
Protein and Peptide Analysis by LC-MS: Experimental Strategies
Edited by Thomas Letzel

Published by the Royal Society of Chemistry, www.rsc.org

physical stability; some of the attachment bonds are stable under conventional MS fragmentation conditions (*N*-glycosidic bond) and the others are labile (*O*-glycosidic or phosphoester bond), which often requires application of alternative fragmentation methods and strategies. Additionally challenging is the overall low abundance of PTMs. Most of the regulatory PTMs are dynamically attached or removed from the modified protein by specific enzymes; therefore, the PTMs tend to be substoichiometric—at most points in the protein lifetime only a fraction of protein molecules are modified. This in turn requires application of more sensitive analytical tools or higher input amounts than in the conventional protein analysis.

Two main analytical aspects need to be addressed in the analysis of every PTM: elucidation of the modification structure and identification of the attachment position(s) on the protein. The latter implies that PTMs ideally have to be analysed on intact proteins. However, this approach is relatively seldom pursued, mainly due to the fact that protein mixtures are difficult to separate and that intact proteins are difficult to ionize and fragment in a mass spectrometer. An alternative approach, liquid chromatography–mass spectrometry (LC-MS) of peptides has emerged as a method of choice for studying proteins and simpler PTMs on a large scale. In this approach, the protein (or proteome) of interest is digested into peptides, which are then separated on a HPLC column, ionized and fragmented in a mass spectrometer. The peptide sequences are identified from fragmentation spectra by searching against protein databases and then mapped back to protein sequences.[2] In most cases the PTMs stay attached to the peptide backbone during sample preparation and are therefore analysed together with the peptide. There are a couple of analytical challenges in this approach; first, the proportion of modified peptides in the whole proteome digest is usually extremely low and biochemical enrichment protocol(s) need to be applied prior to LC-MS analysis. Secondly, depending on the chemical and physical properties of the analysed amino acid, a proper LC method needs to be chosen. The most commonly used upstream LC methods are reversed-phase (RP),[3] hydrophobic interaction chromatography (HILIC)[4] and strong cation exchange chromatography (SCX), the latter usually used as the first dimension in a 2D-HPLC setup.[5] Thirdly, the optimal peptide fragmentation method for the PTM of interest needs to be applied. Stable modifications can be readily measured by collision-induced dissociation (CID),[6] whereas labile modifications may require a special fragmentation regime (*e.g.* multistage activation, neutral loss-dependent MS^3)[7,8] or other fragmentation methods, such as electron capture- or transfer dissociation (ECD or ETD).[9,10]

The process of choosing of a proper LC-MS method to study specific PTM can therefore be quite complex and there is no universal protocol that can be applied to analysis of all PTMs. The example presented here is a revised workflow for enrichment and analysis of protein Ser/Thr/Tyr phosphorylation[11] using SCX, TiO_2 chromatography and downstream LC-MS measurement that has so far been used in several large-scale phosphoproteomics studies in eukaryotes[12,13] and prokaryotes.[14,15] Parts of this protocol can also be applied to enrichment and analysis of other PTMs; SCX can be used to enrich for N-terminal acetylation,[16] whereas TiO_2 can be used for enrichment of sugars containing negatively charged residues, such as sialic acid.[17]

9.2 Materials

9.2.1 In-Solution Protein Digestion

- Denaturation buffer: 6 M urea (Sigma), 2 M thiourea (Invitrogen), 1% n-octylglucoside (w/v) (Roche) in 10 mM HEPES buffer (Sigma), pH 8.0
- Reduction buffer: 1 M dithiothreitol (Sigma) in 50 mM ammonium bicarbonate (Sigma)
- Alkylation buffer: 550 mM iodoacetamide (Sigma) in 50 mM ammonium bicarbonate
- Protease 1: Lysyl endopeptidase LysC (Waco)
- Protease 2: Trypsin, sequencing grade, modified (Promega)
  - ⓘ All solvents should be prepared with high purity deionized Milli-Q water of the resistivity 18.2 MΩcm (Millipore Q-Gard 2 cartridge). Denaturation, reduction and alkylation buffers can be frozen as stock solutions at $-20\,°C$

9.2.2 Strong Cation Exchange (SCX) Chromatography

- SCX solvent A: 5 mM potassium dihydrogen phosphate (Merck), 30% acetonitrile (Merck); acidify with trifluoroacetic acid (Merck) to pH 2.7
- SCX solvent B: 5 mM potassium dihydrogen phosphate, 30% acetonitrile, 350 mM potassium chloride (Merck), acidify with trifluoroacetic acid to pH 2.7

9.2.3 Titanium Oxide (TiO₂) Chromatography

- Loading solution: $30\,mg\,ml^{-1}$ 2,5 dihydrobenzoic acid (Fluka), 80% acetonitrile in water
- Washing solution I: 30% acetonitrile/3% trifluoroacetic acid
- Washing solution II: 80% acetonitrile/0.1% trifluoroacetic acid
- Elution solution: 40% NH_4OH (aq, 25% NH_3, Fluka), 60% acetonitrile (pH > 10.5)
- Beads: Titansphere TiO_2 (10 μm; GL Sciences)

9.2.4 Liquid Chromatography – Mass Spectrometry (LC-MS)

- HPLC solvent A: 0.5% acetic acid (Fluka) in water
- HPLC solvent B: 0.5% acetic acid, 80% acetonitrile in water
- HPLC loading solvent: 1% trifluoroacetic acid, 2% acetonitrile in water
- Stage tips:[18,19] Empore C8 Disk (Varian)
- HPLC columns: Pulled fused silica PicoTip™ Emitter (New Objective)

- HPLC columns: Reversed-phase material for nano-HPLC column: Reprosil-Pur C18-AQ, 3 μm (Dr. Maisch)
  - ⓘ Acetic acid is the optimal ion pairing reagent for the column packing material used here; an alternative is 0.1% formic acid. Do not use more than 0.025% TFA in LC-MS solvents, as it may interfere with electrospray ionization (higher content TFA can only be added into the sample prior to loading)

9.3　Methods

9.3.1　In-Solution Protein Digestion

- Dissolve the protein sample in denaturation buffer
- Add reduction buffer to the sample to final concentration of 1 mM DTT; incubate 1 h at room temperature
  - ⓘ This protocol is optimized for protein amounts from 2 to 20 mg. During cell lysis and protein extraction, keep the salt content at a minimum since it may interfere with subsequent SCX chromatography. If needed, precipitate proteins with acetone or chloroform/methanol prior to analysis. Do not heat-up the samples during protein solubilization and digestion; high concentration of urea will lead to carbamylation of free amino groups
- Add alkylation buffer to the sample to final concentration of 5.5 mM IAA; incubate 1 h at room temperature in the dark
- Check pH (should be 8.0); adjust if necessary
  - ⓘ If necessary, adjust the pH with a very low volume of 1M Tris-HCl, pH 8.0; note that the total salt concentration should not exceed 10 mM since it may interfere with the SCX chromatography
- Add 1 μg of lysyl endopeptidase LysC per 100 μg protein and incubate for 3 h at room temperature
- Dilute sample with 4 volumes of water
  - ⓘ To keep the salt concentration low, use pure water rather than ammonium bicarbonate buffer to dilute the sample prior to trypsin incubation. Check pH *after* you add trypsin; if necessary, adjust the pH with a very low volume of 1 M Tris-HCl, pH 8.0
- Check pH (should be 8.0); adjust if necessary
- Add 1 μg trypsin per 100 μg sample protein and incubate overnight at room temperature. If only LysC digestion is desired, add 1 μg of LysC instead of trypsin and incubate overnight at room temperature

9.3.2　SCX Chromatography

- Acidify the protein digest to pH 2.7 with trifluoroacetic acid and centrifuge the sample to remove any precipitate that may form
- Load the protein digest onto an equilibrated SCX column (GE Healthcare Resource S, 1 mL) in SCX solvent A at a flow rate of 1 mL/min

- ⓘ During sample loading onto the SCX column it is important to monitor conductivity; under the conditions employed here conductivity higher than 4 mS cm^{-1} will lead to decreased binding of peptides. In that case the flow-through should be diluted with water and reloaded onto the column
- Collect the flow-through
  - ⓘ Collection and subsequent separate analysis of the flow-through is extremely important since multiply phosphorylated peptides will not bind to the SCX column; this fraction is usually the richest in the number of identified phosphorylation sites, especially in the analysis of eukaryotic phosphorylation
- Elute the bound peptides with a linear gradient of 0–30% of SCX solvent B in 30 min at a flow rate of 1 ml min^{-1}; collect 2 mL fractions
  - ⓘ It is recommended to monitor the conductivity throughout the SCX run, as it is the best measure for the gradient stability; under conditions employed, at 30% of the SCX solvent B the conductivity typically reaches 13 mS cm^{-1}. This method can be applied for both, tryptic and LysC peptides
- Wash the SCX column with five column volumes of 100% SCX solvent B

9.3.3 TiO$_2$ Chromatography

- Each SCX fraction, including the flow-through, is incubated with the TiO$_2$ beads separately, and is referred to as 'sample' in the text
- Weigh the TiO$_2$ beads into a dedicated tube; for each sample to be analysed, weigh 5 mg of beads (*e.g.* weigh a total of 50 mg if you want to process 10 SCX fractions)
- Add the loading solution to the sample (1:6)
  - ⓘ This step is optional and should be used to increase the binding specificity in the samples with very little phosphorylation (*e.g.* prokaryotic cell lysates); otherwise it should be avoided because it increases the probability of DHB contamination during LC-MS. Alternatively, it is possible to use lactic acid as a competitive binder instead of DHB.[20]
- Add 100 µL of loading solution to the TiO$_2$ beads; shake for 10 min at room temperature
- Add aliquots containing 5 mg of the TiO$_2$ beads slurry into each sample
- Incubate for at least 30 min (end-over-end rotation) at room temperature; centrifuge and discard supernatant
- Wash with 1.5 mL of washing solution I; shake vigorously; centrifuge and discard supernatant
- Wash with 1.5 ml of washing solution II; shake vigorously; centrifuge and discard supernatant
- Prepare one C-8 microcolumn ('stage-tip') for each sample by placing a ~1 mm^2 piece of Empore C$_8$ material into a 200 µL (yellow) pipette tip, as described previously[18,19]

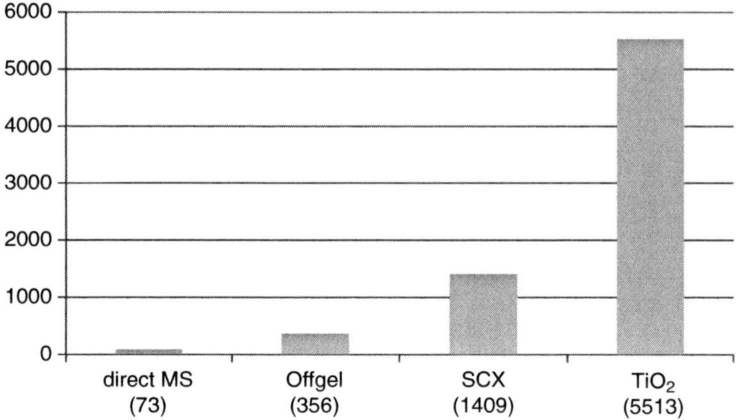

Figure 9.1   Typical enrichment efficiency of phosphopeptides from a HeLa cell digest after SCX and TiO_2 chromatography. Note that the number of detected phosphopeptides is about 75 times higher after TiO_2 enrichment compared to a direct MS measurement of the digest.

- ⓘ C-8 microcolumns can also be commercially obtained (Thermo Fisher Scientific)
- Transfer the TiO_2 beads into dedicated C-8 microcolumns
- Elute each sample with 3×100 µL of elution solution; collect and pool the eluates from each sample. Check pH of the last eluate. It should be >10; if it is not, elute with additional 100 µL of elution solution
  - ⓘ To quickly neutralize the high pH upon elution, it is good to elute the samples into a small volume (~ 50 µL) of the HPLC loading solvent
- Dry the eluates to $\sim$5 µL
  - ⓘ Do not heat the samples during vacuum concentration and make sure they do not run dry. Note that desalting of the samples on either a StageTip or a HPLC pre-column may lead to a loss of hydrophilic phosphopeptides (*e.g.* shorter or multiply phosphorylated peptides)
- Add HPLC loading solvent to 10 µL (final concentration of 1% acetonitrile and $\sim$0.5% trifluoroacetic acid). The samples are now ready for nano-LC-MS

Typical phosphopeptide enrichment efficiency in SCX and TiO_2 is depicted in Figure 9.1.

9.3.4   LC-MS

9.3.4.1   *Liquid Chromatography*

- The liquid chromatography (LC) part of the analytical LC-MS system described here consists of a Thermo EasyLC system (Thermo Fisher Scientific) comprising nanoflow pumps and a thermostated micro-autosampler with a 20 µL injection loop

- Pack an analytical column in a 20 cm fused silica emitter (New Objective, 75 µm inner diameter with a 5–8 µm laser-pulled tip), with a methanol slurry of reversed-phase C-18 resin at a constant helium pressure (50 bar) using a bomb-loader device (Thermo Fisher Scientific), as described previously[21]
  - ⓘ The advantage of the resin used is the fact that it is active at low organic buffer content (*e.g.* less than 2% acetonitrile). In this way the more hydrophobic ion-pairing reagents such as heptafluorobutyric acid (HFBA), which interfere with the electrospray ionization, can be avoided in the HPLC solvents
- Connect the packed emitter (C-18 RP HPLC column) directly to the outlet of the six-port valve of the HPLC autosampler through a 20 cm long, 25 µm inner diameter fused silica transfer line (Composite Metals) and a micro Tee-connector (Upchurch) ('liquid junction' connection)
  - ⓘ Note that there is no pre-column or split in this LC-MS setup
- Load 2–5 µL of the phosphopeptide mixture using the HPLC autosampler onto the packed emitter at a fixed maximum pressure of 280 bar ('Intelliflow' feature) using 2% of HPLC solvent B
- After loading, reduce the flow rate to 200 nL min^{-1} and increase the HPLC solvent B content to 10%
- Separate and elute the bound peptides with a 90 min linear gradient from 10–30% of HPLC solvent B. Wash out hydrophobic peptides by linearly increasing the HPLC solvent B content to 80% over 15 min

9.3.4.2 Mass Spectrometry

- All mass spectrometric experiments discussed here are performed on an LTQ orbitrap 'XL' or 'Classic' mass spectrometer (Thermo Fisher Scientific) connected to an EasyLC nanoflow system (Thermo Fisher Scientific) *via* a nano-electrospray LC-MS interface (Thermo Fisher Scientific)
  - ⓘ Control the timing between the MS and the LC system with a standard double contact closure cable
- Operate the mass spectrometer in the data-dependent mode to automatically switch between MS and MS/MS using the Tune and Xcalibur software package
- Use the following settings in the Tune acquisition software:
  - FT full scan: accumulation target value 1E6; max. fill time 1000 ms
  - FT MS^n: accumulation target value 5E4; max. fill time 500 ms
  - IT MS^n: accumulation target value 5E3; max. fill time 150 ms
- For accurate mass measurements enable the 'Lock mass' option in both MS and MS/MS mode in the Xcalibur software.[22] Use the background polydimethylcyclosiloxane (PCM) ions generated from ambient air (*e.g.* $m/z = 445.120025$) for internal recalibration in real time. For single SIM scan injections of the lock mass ion into the C-trap set the lock mass 'ion gain' at 10% of the target value of the full mass spectrum. If the fragment ion measurements are performed in the Orbitrap, use the PCM ion at m/z 429.088735 (PCM with neutral methane loss)

- In the 'Xcalibur Instrument Setup' create a data-dependent acquisition method in which full scan MS spectra, typically in the m/z range 300–1800, are acquired by the orbitrap detector with resolution $R = 60\,000$ (defined at m/z 400)
- For high-accuracy and full mass range measurements of fragment ions, set the data-dependent MS^2 of the three most intense multiply charged ions ($z \geq 2$) to be measured in the FT analyser (Orbitrap) at the resolution of $R = 15\,000$ and enable the HCD option (higher-energy C-trap dissociation). Set the first mass in mass range to $m/z = 80$. This will allow for low-mass reporter ions such as immonium ions to be identified.[23] Note that the settings may be slightly different for the LTQ Orbitrap Velos[13]
  - ⓘ This gives a total scan cycle time (full scan + 3 MS^2 events) of up to 5 s on an LTQ Orbitrap XL, but only about 2 s on the newer generation LTQ Orbitrap Velos
- For fast scanning and high sensitivity but low resolution measurements of fragment ions, set the data-dependent MS^2 of the five most intense multiply charged ions ($z \geq 2$) to be measured in the linear ion trap. Enable the preview mode for FTMS master scans to perform data-dependent MS^2 *in parallel* with the full scan in the Orbitrap. Set the fragmentation mode to CID and enable the multistage activation fragmentation option by which the neutral loss species at 97.97, 48.99, or 32.66 m/z below the precursor ion will be successively activated for 30 ms each (*pseudo MS^3*).[7] Typical features of a MS^2 spectrum acquired in the linear ion trap using multistage activation are shown in Figure 9.2
  - ⓘ Multistage activation (MSA) produces information-rich spectra, where many of the fragment ions show pronounced neutral loss of phosphoric acid (*e.g.* −97.97, −48.99, or −32.66 for singly-, doubly- or triply-charged fragment ions, respectively). This information is very useful in validation of the phosphopeptide spectra. Note that this feature is useful for analysis of other labile modifications, such as *O*-glycosylation (the MSA values have to be adjusted for the corresponding modification)
- Standard acquisition method settings:
  - Electrospray voltage, 2.4 kV
- Source settings have to be optimized for the emitter and nano-LC-MS setup:
  - No sheath and auxiliary gas flow
  - Ion transfer (heated) capillary temperature, 150 °C
  - Collision gas pressure, 1.3 mTorr
  - Dynamic exclusion of up to 500 precursor ions for 60 s upon MS/MS; exclusion mass width of 10 ppm
  - Normalized collision energy using wide-band activation mode; 35% for both CID and HCD
  - Ion selection thresholds: 1000 counts for CID and 10 000 counts for HCD
  - Activation $q = 0.25$; activation time $= 30$ ms

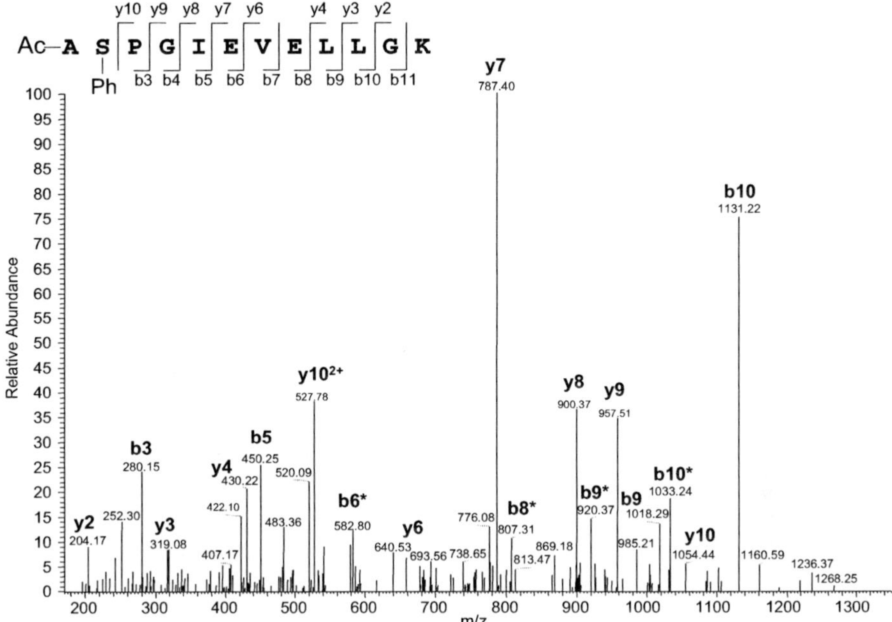

Figure 9.2   A typical MS/MS spectrum of a phosphopeptide obtained in a linear ion trap using multistage activation (MSA). A consequence of MSA is the presence of fragment ions with and without phosphate group attached (*e.g.* b10 and b10*; the mass difference of 98 corresponds to a neutral loss of phosphoric acid from the peptide backbone).

References

1. C. Walsh, *Posttranslational Modification of Proteins: Expanding Nature's Inventory*, Roberts and Co., Englewood, CO, 2006.
2. R. Aebersold and M. Mann, *Nature*, 2003, **422**, 198.
3. M. C. McMaster, *LC/MS: A Practical User's Guide*, John Wiley & Sons, Hoboken, NJ, 2005.
4. A. J. Alpert, *J. Chromatogr.*, 1990, **499**, 177.
5. D. A. Wolters, M. P. Washburn and J. R. Yates, 3rd, *Anal. Chem.*, 2001, **73**, 5683.
6. J. M. Wells and S. A. McLuckey, *Methods Enzymol.*, 2005, **402**, 148.
7. M. J. Schroeder, J. Shabanowitz, J. C. Schwartz, D. F. Hunt and J. J. Coon, *Anal. Chem.*, 2004, **76**, 3590.
8. S. A. Beausoleil, M. Jedrychowski, D. Schwartz, J. E. Elias, J. Villen, J. Li, M. A. Cohn, L. C. Cantley and S. P. Gygi, *Proc. Natl. Acad. Sci. U S A*, 2004, **101**, 12130.
9. R. A. Zubarev, D. M. Horn, E. K. Fridriksson, N. L. Kelleher, N. A. Kruger, M. A. Lewis, B. K. Carpenter and F. W. McLafferty, *Anal. Chem.*, 2000, **72**, 563.

10. J. J. Coon, J. Shabanowitz, D. F. Hunt and J. E. Syka, *J. Am. Soc. Mass Spectrom.*, 2005, **16**, 880.
11. J. V. Olsen and B. Macek, *Methods Mol. Biol.*, 2009, **492**, 131.
12. J. V. Olsen, B. Blagoev, F. Gnad, B. Macek, C. Kumar, P. Mortensen and M. Mann, *Cell*, 2006, **127**, 635.
13. J. V. Olsen, J. C. Schwartz, J. Griep-Raming, M. L. Nielsen, E. Damoc, E. Denisov, O. Lange, P. Remes, D. Taylor, M. Splendore, E. R. Wouters, M. Senko, A. Makarov, M. Mann and S. Horning, *Mol. Cell Proteomics*, 2009, **8**, 2759.
14. B. Macek, F. Gnad, B. Soufi, C. Kumar, J. V. Olsen, I. Mijakovic and M. Mann, *Mol. Cell Proteomics*, 2008, **7**, 299.
15. B. Macek, I. Mijakovic, J. V. Olsen, F. Gnad, C. Kumar, P. R. Jensen and M. Mann, *Mol. Cell Proteomics*, 2007, **6**, 697.
16. A. O. Helbig, S. Gauci, R. Raijmakers, B. van Breukelen, M. Slijper, S. Mohammed and A. J. Heck, *Mol. Cell Proteomics*, **9**, 928.
17. M. R. Larsen, S. S. Jensen, L. A. Jakobsen and N. H. Heegaard, *Mol. Cell Proteomics*, 2007, **6**, 1778.
18. J. Rappsilber, Y. Ishihama and M. Mann, *Anal. Chem.*, 2003, **75**, 663.
19. J. Rappsilber, M. Mann and Y. Ishihama, *Nat. Protoc.*, 2007, **2**, 1896.
20. Y. Ishihama, J. Rappsilber, J. S. Andersen and M. Mann, *J. Chromatogr., A*, 2002, **979**, 233.
21. J. V. Olsen, L. M. de Godoy, G. Li, B. Macek, P. Mortensen, R. Pesch, A. Makarov, O. Lange, S. Horning and M. Mann, *Mol. Cell Proteomics*, 2005, **4**, 2010.
22. J. V. Olsen, B. Macek, O. Lange, A. Makarov, S. Horning and M. Mann, *Nat. Methods*, 2007, **4**, 709.
23. N. Sugiyama, T. Masuda, K. Shinoda, A. Nakamura, M. Tomita and Y. Ishihama, *Mol. Cell Proteomics*, 2007, **6**, 1103.

CHAPTER 10

LC-MS for the Determination of the Enzymatic Activity of Proteins

ROMY K. SCHEERLE AND JOHANNA GRAßMANN

Institute for Chemical-Technical Analysis and Chemical Food Technology, Center of Life and Food Sciences Weihenstephan, Technische Universität München, Weihenstephaner Steig 23, 85350 Freising, Germany

10.1 Introduction

Methods for studying enzymatic reactions are manifold, such as spectroscopic, electrochemical or radiometric assays. Spectroscopic and mainly photometric measurements are the most frequent employed methods in enzymatic analysis. This is due to the easy handling, moderate costs and the accuracy of the methods. However, these methods require labelling of the substrates, e.g. chromogenic or fluorogenic substrates or radioisotope labelled substances.[1] These labelled substrates may, however, affect the reaction rate and/or kinetics of the enzymatic reaction.[2] The use of natural substrates is therefore desirable and can be achieved by utilizing mass spectrometric (MS)-based enzymatic assays. Besides the advantage of the applicability of natural substrates, MS-based assays offer the possibility to observe not only the product of an enzymatic reaction but also the substrates and in some cases also the enzyme and enzyme complexes (for latter see also Chapter 11). Many applications have been described in recent years to follow enzymatic reactions by LC-MS. In most of these applications the enzymatic reaction is conducted in either a single reaction tube or a well plate.

RSC Chromatography Monographs No. 15
Protein and Peptide Analysis by LC-MS: Experimental Strategies
Edited by Thomas Letzel
© The Royal Society of Chemistry 2011
Published by the Royal Society of Chemistry, www.rsc.org

After defined time points the reactions are quenched and the resulting aliquots are subjected to (LC-)MS analysis. This setup allows obtaining information about the concentration of substrate and product(s) at defined time points. A different approach is the real-time monitoring of enzymatic reactions with direct infusion of the assay mixture into the mass spectrometer using either a syringe pump or a robot system. Using this continuous flow measurement, a real-time kinetic of the enzymatic reaction can be obtained.[3]

The differences of these three setups as well as their advantages and disadvantages will be discussed in the present chapter. The enzymatic reaction of lysozyme with a hexasaccharide will serve as an example to illustrate several aspects of the various approaches.

10.2   Materials

10.2.1   The Example Assay

Hen egg white lysozyme (HEWL, muramidase) is a well-known protein that catalyses the hydrolysis of β-1,4-glycosidic linkages in certain Gram-positive bacterial cell walls and chitin.[4–6] Therefore the hexasaccharide hexa-*N*-acetyl-chitohexaose ((GlcNAc)$_6$) can be used as substrate. Lysozyme cleaves (GlcNAc)$_6$ into (GlcNAc)$_2$ and (GlcNAc)$_4$ as major products.[7] The assay conditions were adopted from Dennhart and Letzel.[8]

10.2.2   Chemicals

Ammonium acetate (NH$_4$Ac, M$_r$ 77.08, > 98%) and acetic acid were purchased from Merck (Darmstadt, Germany); acetonitrile (M$_r$ 41.05, > 99.9%) and hen egg white lysozyme (HEWL, 95%; M$_r$ 14.3 kDa, 50 000 units/mg protein) were purchased from Sigma-Aldrich (Steinheim, Germany). Hexa-*N*-acetyl-chitohexaose ((GlcNAc)$_6$, M$_r$ 1237.17, ≥95%) was obtained from Seikagaku Biobusiness (Tokyo, Japan). All buffers and solutions were prepared in LC-MS reagent water from J.T.Baker (Deveter, Holland).

10.2.3   Mass Spectrometer

- TripleQuadrupole LC/MS 6410 instrument, Agilent Technologies, Waldbronn, Germany
- MassHunter Workstation software, Qualitative Analysis, Agilent Technologies, Waldbronn, Germany

10.2.4   LC-MS Method

- Assay buffer: 10 mM NH$_4$Ac pH 5.2
- Substrate solution: 50 μM (GlcNAc)$_6$ in 10 mM NH$_4$Ac pH 5.2
- Enzyme solution: 10 μM lysozyme in 10 mM NH$_4$Ac pH 5.2
- Stopping reagent: 90% acetonitrile

- HPLC: 1200 Series, Agilent Technologies, Waldbronn, Germany
- HPLC solvents: 90% acetonitrile
- Column: LichroCART, 100 Diol, 5 μm, Merck, Darmstadt, Germany

10.2.5 Direct Infusion

- Assay buffer: 10 mM NH_4Ac pH 5.2
- Substrate solution: 50 μM $(GlcNAc)_6$ in 10 mM NH_4Ac pH 5.2
- Enzyme solution: 10 μM lysozyme in 10 mM NH_4Ac pH 5.2
- Syringe: 100 μL, Hamilton-Bonaduz, Switzerland
- Syringe pump: Model 11 Plus, Harvard Apparatus, Hugo Sachs Elektronik, Hugstetten, Germany

10.2.6 Robot Infusion

The TriVersa NanoMate® system is an automated nano-electrospray ion source for mass spectrometers that enables the direct infusion of samples from well plates to the MS. The system consists of a robot part and an ESI chip, consisting of 20×20 nozzles, through which the samples are delivered. The nano-electrospray is generated by applying a voltage and head pressure to the sample in the pipette tip.

- Assay buffer: 10 mM NH_4Ac pH 5.2
- Substrate solution: 20 μM $(GlcNAc)_6$ in 10 mM NH_4Ac pH 5.2
- Enzyme solution: 4 μM lysozyme in 10 mM NH_4Ac pH 5.2
- Robot device: Instrument Advion TriVersa NanoMate©, Advion Bio-Sciences, Ithaca, NY
- Conductive pipette tips: Sample tip rack, 384, Advion BioSciences, Ithaca, NY
- ESI chip with 400 nozzles: HD_A_384, Advion BioSciences, Ithaca, NY
- Well plates: Thermo-Fast® 96 skirted, low profile, Thermo Scientific, Surrey, UK

10.3 Methods

10.3.1 LC-MS

10.3.1.1 Enzymatic Assay

- Mix appropriate volumes of enzyme and substrate solutions in a reaction tube to achieve the desired final concentrations
    - ➤ Example assay: Mix equal volumes of enzyme and substrate solutions to achieve final assay concentrations of 25 μM $(GlcNAc)_6$ and 5 μM lysozyme
    - ⓘ In general the mixing order is irrelevant, but keep the same order of mixing in all experiments

- Incubate the mixture, if possible at the temperature optimum of the enzyme
  - ⓘ When conducting repeated assays, be aware that the incubation takes place at the same, defined temperature for each of the assays
- At defined time points (*e.g.* 2, 4, 6, 8 min and so on) add the stopping reagent
  - ➢ Example assay: Stop the reaction with 90% acetonitrile
  - ⓘ Depending on your enzyme and your HPLC separation method you may stop the enzyme reaction with acid, alkali or some kind of organic solvent (methanol, acetonitrile)
- Centrifuge or filter the solution

10.3.1.2   *LC-MS Measurement*

- Separate the mixture on HPLC. Typical flow rates for MS detection are 0.1–0.2 mL min^{-1}; if you have to apply higher flow rates you should split the flow before subjecting it to the mass spectrometer
  - ➢ Example assay: Inject 10 µL of the mixture and perform the separation on a LichroCART, 100 Diol, 5 µm column with 90% acetonitrile (0.1% formic acid)/10% H$_2$O. In our case, however, we observed hydrolysis of the substances on the column
  - ⓘ If you do not use a 'make up' flow after the column, you have to use a suitable additive (*e.g.* formic acid for the positive ionization mode or ammonium hydroxide for the negative ionization mode) to achieve sufficient ionization of your molecules

10.3.1.3   *Mass Spectrometric Detection*

- Select the appropriate mass range to detect your substrate and product. The parameters of the ion source and the mass spectrometer depend on the device used. Typical values are: capillary voltage 4–4.5 kV; nebulizer pressure 15–30 psi; drying gas flow rate 5–10 L min^{-1}; drying gas temperature 250–350 °C
  - ➢ Example assay: Set the m/z range at 400–900; this allows the detection of the substrate (GlcNAc)$_6$ ($m/z = 619.3$) and the two products (GlcNAc)$_4$ ($m/z = 831.3$) and (GlcNAc)$_2$ ($m/z = 425.2$). The MS parameters are as follows: drying gas temperature 300 °C; drying gas flow rate 6 L min^{-1}; nebulizer pressure 30 psi; capillary voltage 2.5 kV; positive ionization mode
  - ⓘ In general for LC-MS measurements with flow rates of up to 0.2 mL min^{-1} you may need higher drying gas parameters than for the direct infusion measurements with flow rates of about 5 mL min^{-1}

10.3.2   Direct Infusion or Robot Infusion Measurement

10.3.2.1   *Direct Infusion Enzymatic Assay*

- Mix appropriate volumes of enzyme and substrate solutions in a reaction tube to achieve the desired final concentrations

> Example assay: Mix equal volumes of enzyme and substrate solutions to achieve final assay concentrations of 25 μM (GlcNAc)$_6$ and 5 μM lysozyme
> ⓘ In general the mixing order is irrelevant, but keep the same order of mixing in all experiments
- Transfer the mixture into a syringe, put the syringe into the syringe pump and connect it *via* needle port and PEEK tubing with the inlet of the mass spectrometer source
  > ⓘ Try to conduct these steps quickly, since the enzymatic reaction starts once the enzyme and substrate are mixed

10.3.2.2   Robot Infusion Enzymatic Assay

- Preload two wells of the microplate with the enzyme and substrate solutions
- Program the robot to conduct the following steps:
  1. Aspirating a defined volume of the substrate solution, followed by 1 μL of air and then dispensing the substrate solution in an unused well (in the following referred to as the 'reaction well')
  2. Aspirating a defined volume of the enzyme solution, followed by 1 μL of air and then dispensing the enzyme solution in the reaction well
  3. Mixing the substrate and enzyme solutions in the reaction well
  > Example assay: 10 μL of substrate solution are mixed with 10 μL of enzyme solution (final concentration of 10 μM (GlcNAc)$_6$ and 2 μM lysozyme), the mixing step with a mixing volume of 20 μL is repeated three times
  > ⓘ Compared to the syringe pump in case of the ESI chip device lower concentrations of enzyme and substrate are used in order to avoid clogging of the nozzles
  4. Transport of the mixed solutions (*i.e.* the enzymatic assay) to the ESI chip and initiation of the nano-electrospray with the appropriate values of voltage and gas pressure
  > Example assay: the applied values are 1.8 kV voltage and 0.8 psi head pressure
  > ⓘ In general, due to their higher surface tension aqueous solutions require higher values of voltage and gas pressure than organic solutions.[9] For each single assay you have to find the optimal combination of voltage and pressure in preliminary experiments

10.3.2.3   Mass Spectrometric Detection

- Select an appropriate time range to follow the enzymatic reaction
  > Example assay: Follow the reaction for 20 min; after this time the reaction is completed
- Select the appropriate mass range to detect your substrate and product. The parameters of the ion source and the mass spectrometer depend on the

device used. Typical values for direct infusion are: capillary voltage 4–4.5 kV; nebulizer pressure 15–30 psi; drying gas flow rate 5–10 L min^{-1}; drying gas temperature 250–350 °C. Values for the robot infusion are given in the example assay

> Example assay: Set the *m/z* range at 400–900; this allows the detection of the substrate (GlcNAc)$_6$ (*m/z* = 619.3) and the two products (GlcNAc)$_4$ (*m/z* = 831.3) and (GlcNAc)$_2$ (*m/z* = 425.2). The measurements are performed in the positive ionization mode. The parameters for the direct infusion are as follows: 250 °C drying gas temperature; 6 L min^{-1} drying gas flow; 15 psi nebulizer pressure and 4 kV capillary voltage. The MS parameters for the robot infusion are 170 °C drying gas temperature and 1.2 L min^{-1} drying gas flow. No nebulizer is used and capillary voltage must be set to 0 V

ⓘ In general, for LC-MS measurements with flow rates of up to 0.2 mL min you may need higher drying gas parameters than for the direct infusion measurements (flow rates of about 5 μL min^{-1}) and for the robot infusion measurements (flow rates of 100–200 nL min^{-1})

10.3.3 Data Analysis

Extract the ion chromatograms of substrate and product(s). For LC-MS, integrate the respective peaks for each time point and plot the areas obtained against the time points. For the direct infusion measurements you will obtain a continuous ion chromatogram. Figure 10.1 schematically illustrates the resulting graphs from a LC-MS and a direct infusion measurement.

In case of sodium and/or potassium adducts you have to sum the respective areas or traces.

> Example assay: The extracted ion chromatogram signals were summed for the following compounds: for the products (GlcNAc)$_2$ and (GlcNAc)$_4$ the signals of [(GlcNAc)$_n$ + H$^+$] and [(GlcNAc)$_n$ + Na$^+$] were summed and for the substrate (GlcNAc)$_6$ only the signal [GlcNAc)$_6$ + 2H$^+$] was considered

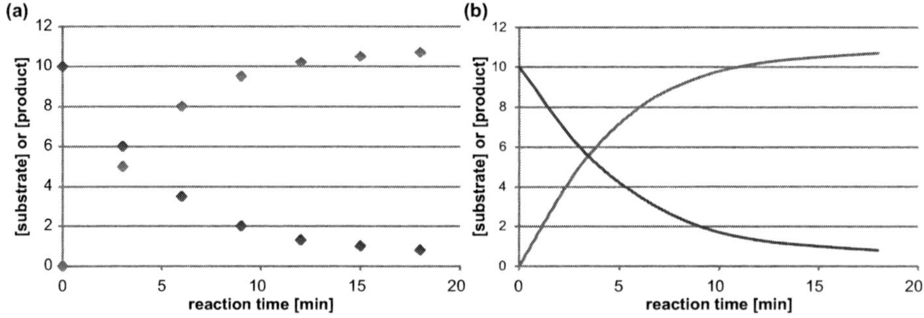

Figure 10.1   Change in substrate and product intensities of an enzymatic assay: (a) graph of LC/MS results; (b) graph of a direct infusion measurement.

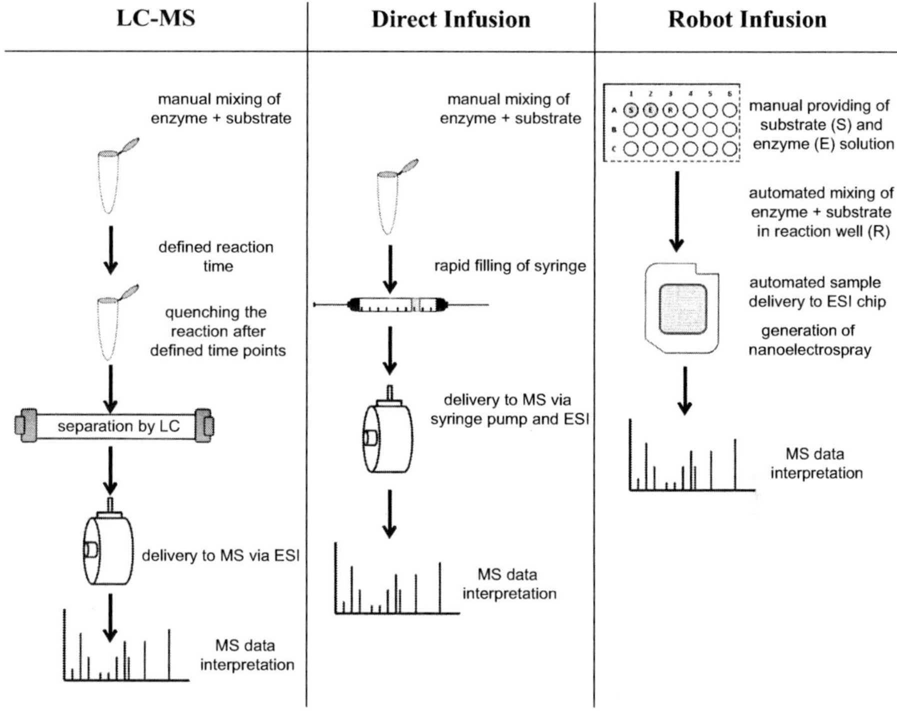

Figure 10.2 Procedural method of the different setups for studying enzymatic reactions.

10.3.4 Comparison of LC-MS, Direct Infusion and Robot Infusion

The LC-MS, direct infusion and robot infusion methods feature several differences. The procedure for each method is shown schematically in Figure 10.2. Additionally, Table 10.1 illustrates the advantages and disadvantages of the methods.

10.4 Notes

10.4.1 General Remarks for Working with Enzymes

10.4.1.1 Enzyme Stability

- Most enzymes are sensitive to temperature, but freezing enzymes may sometimes cause enzyme inactivation and/or denaturation. In this case, freezing should be avoided and enzyme should be kept on ice or at +4 °C
- pH may also affect the stability of enzymes
- Many enzymes are inhibited by heavy metals, so you should take care to exclude them from the assay

Table 10.1 Overview of advantages and disadvantages of setups for the study of enzymatic reactions

LC-MS	*Direct infusion*	*Robot infusion*
Advantages		
Buffer must not be MS-compatible	Real time measurement of the enzymatic reaction	Real-time measurement of the enzymatic reaction
Due to separation of the reaction mixture interferences between the metabolites are reduced	Smaller sample volumes are sufficient	Even smaller sample volumes than for the syringe pump are sufficient
Complex matrices can be used, *e.g.* plasma samples; interfering substances may be separated from the reaction mixture prior to the LC-MS step	No sample preparation necessary	Less delay between mixing and first data acquisition due to automation
		No sample preparation necessary
		No cross contamination since for each measurement a single pipette tip and nozzle are used
		No purging necessary after the measurements
Disadvantages		
Method development for LC separation may be time consuming	Delay between mixing and first data acquisition due to mixing and filling of the syringe	MS-compatible buffer must be used
Sample preparation is time consuming	MS-compatible buffer must be used	Metabolites of the reaction mixture may interfere
Substances may stick in the tubing	Metabolites of the reaction mixture may interfere	Metabolites may interact with the pipette tip or the chip
Only single time points can be measured	Syringe and tubing must be purged exhaustively after each measurement to avoid cross contamination	
The reaction mixture may undergo undesirable reactions on the column, *e.g.* hydrolysis	Substances may stick in the tubing	

- Some enzymes are sensitive to vigorous mixing, thus too strong mixing should be avoided
- Ultrasonics may also affect enzymatic activity (enhance or decrease it)

10.4.2   General Remarks for Mass Spectrometric Applications

- It is strongly recommended to utilize solvents of the highest purity grade (LC/MS grade) and to filter all solutions prior to measurement
- For filtering, make sure that the filter is compatible with your solvent and does not bind proteins
- Avoid high sodium or potassium concentrations to prevent sodium and/or potassium adducts. For this reason when adjusting the pH of buffers use ammoniac or ammonium hydroxide and formic or acetic acid solutions, respectively

10.4.3 Remarks Regarding MS-Based Enzymatic Assays

10.4.3.1 Selection of a Suitable Substrate

- Substrate has a characteristic m/z and is detectable *via* MS
- Enzyme changes the substrate in a characteristically way that is detectable *via* MS (not too small fragments; they should have $m/z \geq 100$)
- Substrate and product m/z values do not overlap
- Choose the appropriate enzyme and substrate concentrations
- In case of more complex assays which may require several cofactors, salts and other additives, mix all required solutions and start the reaction by adding the enzyme

10.4.3.2 Selection of a Suitable Buffer

- Buffer and pH must be in accordance with the requirements of the chosen enzyme (pH range, salt concentration)
- In case of direct infusion measurement, the MS-compatibility of your buffer is an additional criterion: utilize volatile salts like acetate and dependent on your detection method (positive or negative mode), a suitable pH that leads to sufficient ionization of your molecules. However, also take care about the optimum pH for your enzyme (see above)
- Before starting the measurements, make a complete digest of your substrate to check the activity of the enzyme and to check that you have chosen the right substrate

Another continuous-flow technique for the study of protein functions (*e.g.* enzymatic activity) can be found in Chapter 11.

References

1. A. Liesener and U. Karst, *Anal. Bioanal. Chem.*, 2005, **382**, 1451.
2. T. Letzel, E. Sahmel-Schneider, K. Skriver, T. Ohnuma, and T. Fukamizo, *Carbohydr. Res.* 2011, **346**, 863.
3. R. Scheerle, J. Graßmann and T. Letzel, *Anal. Meth.*, 2011, doi:10.1039/C0AY00727G.
4. R. A. John, in *Enzyme Assays: A Practical Approach*, 2nd edn, eds. R. Eisenthal and M. J. Danson, Oxford University Press, Oxford, UK, 2002.
5. B. W. Matthews, *FASEB J.*, 1996, **10**, 35.
6. A. Rau, T. Hogg, R. Marquardt and R. Hilgenfeld, *J. Biol. Chem.*, 2001, **276**, 31994.
7. T. Fukamizo, *Curr. Protein Pept. Sci.*, 2000, **1**, 105.
8. N. Dennhart and T. Letzel, *Anal. Bioanal. Chem.*, 2006, **386**, 689.
9. C. Van Pelt, S. Zhang and J. Henion, *J. Biomol. Tech.*, 2002, **13**, 72.

Functional Analysis of Proteins, Including LC-MS and Special Freeware

MICHAEL KRAPPMANN[1] AND THOMAS LETZEL[2]

[1] Institut für Forschung und Weiterbildung, University of Applied Sciences, Hochschule Weihenstephan – Triesdorf, Am Hofgarten 4, 85350 Freising, Germany; [2] Competence Pool Weihenstephan (CPW), Technische Universität München, Weihenstephaner Steig 23, 85350 Freising, Germany

11.1 Introduction

Proteins have a function in organisms when they are acting as an enzyme (catalyst), when they are regulating as complexes (covalent/non-covalent) and/or when they are transporting compounds *via* non-covalent complexes. For this reason analytical methods often are applied that need labelled molecules to obtain signals in the detection systems. Mass spectrometry does not need a labelling group, if the analyte can be transferred in the mass spectrometer as an ion (see also Chapter 2).[1,2] Thus, mass spectrometry can be used to detect protein function indirectly *via* post-translational modifications (see also Chapter 9) or directly by monitoring enzymatic conversion reactions (see Chapter 10) or complex formation (see this chapter).

New strategies in coupling liquid flow techniques with mass spectrometry result in protein reaction or interaction in continuous-flow or continuous-flow mixing assays. Chapter 10 gives an example of a continuous flow–mass spectrometry setup for the detection of enzyme kinetics using hen egg white lysozyme (HEWL) and

RSC Chromatography Monographs No. 15
Protein and Peptide Analysis by LC-MS: Experimental Strategies
Edited by Thomas Letzel

Published by the Royal Society of Chemistry, www.rsc.org

substrate. The same setup can be used for the monitoring of noncovalent complexes of HEWL with inhibitors. However, such studies favour continuous-flow mixing assays, as presented in this chapter. Injections of molecules that can inhibit enzymes, especially by forming specific non-covalent protein–ligand complexes, can sometimes be detected by increasing mass to charge ratios of enzyme complexes and/or negative mass to charge (m/z) ratios of product. Such an experiment, and special free software to work out the data, are presented in this chapter.

11.2 Materials and Systems

11.2.1 Chemicals

- Enzyme: Hen egg white lysozyme (HEWL, M_r 14.3 kDa, $\sim$95%, $\sim$50 000 units protein mg)
- Substrate: Hexa-N-acetylchitohexaose (($GlcNAc)_6$), M_r 1237.2, $\geq$95%)
- Inhibitor: N, N′, N′-triacetylchitotriose (($GlcNAc)_3$), M_r 627.6, $\geq$95%)
- Internal standard: Malantide (IS, M_r 1633.9, $\geq$97%)
  - Enzyme, substrate, inhibitor and internal standard were obtained from Sigma-Aldrich (Steinheim, Germany)
- Further chemicals: Ammonium acetate (NH_4Ac, >98%) and acetic acid were purchased from Merck (Darmstadt, Germany), acetonitrile (>99.9%) and methanol (MS grade) from Riedel-de Haën (Seelze, Germany). High purity water was from a Milli-Q system (Millipore, Eschborn, Germany)
  - ⓘ The higher the purity of the applied compounds, the lower the impurity signals in the obtained mass spectra

11.2.2 The Example Enzymatic Assay

- Enzyme HEWL and substrate (($GlcNAc)_6$): see also Chapter 10
- Inhibitor: N, N′, N′-triacetylchitotriose (($GlcNAc)_3$)
  - ⓘ An inhibitor is a regulator (ligand, molecule) that decreases the rate of an enzyme-catalysed reaction if it is present in the reaction mixture[3,4]

11.2.3 HPLC (Sample Introduction)

- HPLC: 1100 Series, Agilent Technologies, Waldbronn, Germany with binary pump, isocratic pump, degasser, well-plate-sampler, column oven and multiwavelength detector
- Software: ChemStation (Rev.B.01.01), Agilent Technologies, Waldbronn, Germany

11.2.4 Mass Spectrometer

- Mass spectrometer: Q-TOF Ultima API, Waters-Micromass, Manchester, UK

- Desolvation gas and cone gas was Nitrogen (purity 4.0) obtained from a nitrogen generator (Domnick Hunter GmbH, Krefeld, Germany) and the collision gas was argon (purity 5.0)

The setting parameters were optimized for the ESI interface (see Chapter 2 for a similar optimization) and final parameters are like the following:

- The measurements were carried out with 80 °C source temperature, 150 °C desolvation temperature, 300 L h^{-1} desolvation gas flow and 50 L h^{-1} cone gas flow
- The mass spectrometric detection was performed in positive detection mode. The capillary voltage was 3000 V, the cone voltage 40 V and the collision energy 10. The mass-range was set to m/z range of 50–3000 with 1 s scan time
  - ⓘ The Q-TOF calibration was done by using NaI and CsI solution. The salt cluster can very well be used for mass spectrometric calibrations in the region of higher mass to charge ratio and to cover a large mass range

11.2.5 Complex Study with Flow Injection Analysis (FIA)

- Assay solution: 90% 10 mM NH_4Ac containing aqueous solution (pH 5.2) with 10% methanol
- Substrate solution: 50 µM $(GlcNAc)_6$ in 10 mM NH_4Ac solution (pH 5.2)
- Enzyme solution: 20 µM HEWL + 4 µM malantide in 10 mM NH_4Ac solution (pH 5.2)
- Inhibitor solution: 1–40 nM in 10 mM NH_4Ac solution (pH 5.2)
- Stopping reagent: 90% acetonitrile
- Syringe pump: 11 Plus, Harvard Apparatus, Hugo Sachs Elektronik, Hugstetten, Germany
- Superloops: Volume: 10 mL, Amersham Biosciences, Pittsburgh, USA
- Tubing: PEEK (polyetheretherketone), 0.17 mm × 350 mm and 0.25 mm × 830 mm, Chromatographie Handel Müller, Fridolfing, Germany
- Mixer: 1/16″PEEK mixing-T, 10 µm UHMWPE frit, Chromatographie Handel Müller, Fridolfing, Germany

11.2.6 Analysis Software

- MassLynx Software, Version 4.0, Waters-Micromass, Manchester, UK
- Achroma Software Tool, University of Applied Science Weihenstephan-Triesdorf, Freising, Germany[5]

11.3 Methods

A typical setup for an optimal performance of complex formation experiments is shown in Figure 11.1. The original setup was developed to obtain a

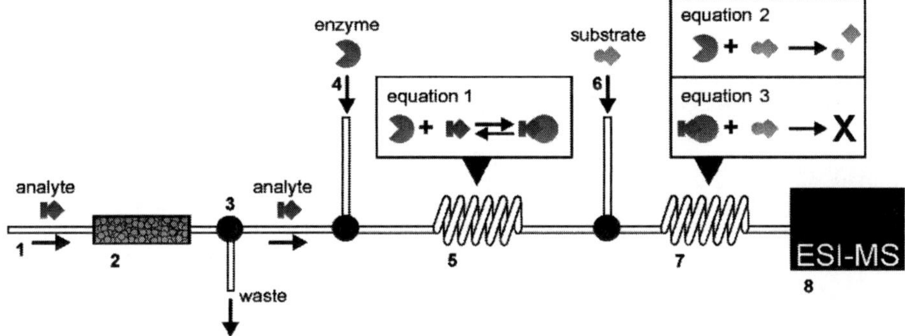

Figure 11.1　Scheme of the on-line continuous-flow mixing system: (1) sample introduction *via* the assay solution +5% methanol; (2) HPLC column; (3) 1:3 flow splitter, 50 µL/min bioassay/150 µL/min waste; (4) enzyme solution; (5) reaction coil A; (6) substrate solution; (7) reaction coil B; (8) mass spectrometer. Enzyme continuously converts the substrate into products (equation 2) if no inhibitor is eluting from the column. Bioactive compounds present in the eluate bind to the enzyme (equation 1), resulting in a decrease of product turnover (equation 3).[6] Reprinted with permission from A.R. de Boer *et al.*: On-line coupling of high-performance liquid chromatography to a continous-flow enzyme assay based on electrospray ionization mass spectrometry. *Anal. Chem.*, **76**, 2004; 3155–161. Copyright 2004 American Chemical Society.

continuous-flow mixing system for the monitoring of enzymatic inhibitors in complex solvents extracted from plants *via* substrate and product ion traces in MS.[6]

The first step is a chromatographic separation leading to a differentiation of single molecules from a complex mixture (parts 1 and 2 in Figure 11.1). The HPLC column eluting compounds (see part 3 in Figure 11.1) is used further in the continuous-mixing flow setup.

- ⓘ Take care that the HPLC mobile phase contains as little organic solvent as possible, because the higher the content the lower the enzymatic activity (due to protein denaturation and compound dissociation)
- ⓘ On the other hand, organic solvents protect proteins from being adsorbed on capillary surfaces and support an efficient ion transfer out of the interface spray into the mass spectrometer (see also Chapter 2). Thus effective detection often requires the addition of a 'make-up' flow after eluting the reaction system prior to entering the mass spectrometer (*e.g.* behind coil 7 and prior to MS 8 in Figure 11.1)

Classical reversed-phase (RP)-HPLC can be applied if the organic content is low enough and no protein denaturation occurs. This is contrary to the requirements of a typical HPLC separation, whereby non-polar compounds elute at high organic content. Techniques like high-temperature (HT)-HPLC are often used to overcome this weakness. The effect of decreasing polarity by increasing the temperature of water and water/organic solvent mixtures thereby plays an extensive role. A detailed description of HT-HPLC can be found in a recent RSC monograph.[7] As a consequence, solvent gradients can partially be substituted by temperature gradients, with the result that lower quantities of organic solvents are needed.

Back to the continuous-mixing flow system; the elution flow out of the separation column (see part 3 in Figure 11.1) containing the separated molecules is mixed with the enzyme-containing flow (see part 4 in Figure 11.1).

A separated compound that elutes and has enzyme-regulating properties is interacting with the enzyme (in coil 5 in Figure 11.1) and influences the enzymatic activity, *i.e.* the enzyme is activated or inactivated. In a second coil (coil 7 in Figure 11.1) this stream is mixed with the substrate solution. Now, an up- or down-regulated substrate conversion compared to the regular situation can be monitored. As long as there is no regulating molecule interacting with the enzyme, a steady amount of substrate will be converted and the chromatographic result for the substrate as well as the product ion trace (as an extracted ion chromatogram, EIC) is a straight line. Inhibiting compounds lead to a negative tip and activating compounds to a positive tip in the resulting product trace. Several compounds—substrate, products and other ionic compounds— can be directly monitored and presented as EICs of the relevant m/z ranges.[8]

In this setup the product formation *versus* its change directly marks the combination of a time point of an HPLC separation with biological activity, and thus leads directly to molecular masses and to hydrophobicity of the regulating compounds. The same is true for the monitoring of non-covalent complexes. The experiment explained below shows the development of non-covalent enzyme–inhibitor complexes through the EIC of the relevant m/z ranges and the use of the Achroma software tool.

11.3.1 Studying Complex Formation (and Simultaneous Reaction Inhibition) with Continuous-Flow Mixing Analysis

The experimental setup in this chapter is a similar on-line system as described above (from de Boer *et al.* in 2004).[6] The continuous-flow mixing system consists of a sample introduction part (see parts 1–3 in Figure 11.1), an enzymatic assay (see parts 4–7 in Figure 11.1) and an ESI-MS (see part 8 in Figure 11.1). The HPLC sample introduction part is substituted by an auto injector for studying the complex formation.

11.3.1.1 *Enzymatic Assay and Inhibition (Complex Formation) Study*

- First, the inhibitor (complex-forming compound) is injected into the system by an autosampler at periodic time intervals (sample introduction part). In the reaction coil (see part 5 in Figure 11.1) inhibitor and enzyme form complexes (see equation 1 in Figure 11.1) for a defined interaction time

- In this way enzyme (see part 4 in Figure 11.1) and substrate (see part 6 in Figure 11.1) are pumped continuously from the two containing superloops into the flow system
- In the reaction coil (see part 7 in Figure 11.1) enzyme–substrate complexes can be formed at times when no inhibitor is injected. During the inhibitor injection periods, both inhibitor–enzyme complexes and substrate–enzyme complexes can be formed in competition. The equilibrium therefor depends thereby on the inhibitor concentration and affinity (see equations 2 and 3 in Figure 11.1)
- The continuous-flow system ends in an ESI-MS, which detects the relevant m/z ranges by time for enzyme, inhibitor, substrate, product, and possible complexes
    - ➢ Example assay: Injection volume of inhibitor is 10 μL (injection rate 4 μL/min, ejection rate 10 μL/min) *via* autosampler. The flow rate of the pump is 20 μL/min. 10 μL of various inhibitor concentrations (see Table 11.1) are injected with 4 min runtime before the next injection. Each injection takes 2.5–3.5 min. The first tubing is used as interaction coil (see part 5 in Figure 11.1) for enzyme–inhibitor interaction with a length of 350 mm and a diameter of 0.17 mm. Thus the interaction time between enzyme and inhibitor is 13 s (0.21 min)
    - ⓘ The reaction time ($r^2\pi{*}L/fr$) of the mixtures is determined by the tubing length L, inner diameter ($2r$) of the tubing and the flow rate (fr) of the solutions and can be adapted by varying these parameters
- Enzyme (HEWL) and substrate ($GlcNAc_6$) are continuously pumped by the HPLC binary pump or two single pumps with a flow rate of 20 μL/min each from a superloop container into the flow system
    - ⓘ The tubing of one pump head of the binary pump is unscrewed from the gradient mixer and is directly connected to one superloop (if possible, use an external pulse damper). The other flow is guided through the HPLC system and then connected with the second superloop
    - ⓘ Cool the superloops down if you store large volumes of enzyme/substrate for an extended continuous-flow enzymatic assay (with ice, if

Table 11.1 Injection concentrations and absolute amount injected of the inhibitor (($GlcNAc)_3$) into the enzyme containing assay with final stoichiometry at each flow-injection analysis

Concentration of inhibitor (μmol L^{-1})	*Abundance of ligand (inhibitor) (nmol)*	*Ratio ($GlcNAc)_3$:($GlcNAc)_6$:HEWL (inhibitor:substrate:enzyme)*
0.1	1	2.5:5:1
0.2	2	5:5:1
0.5	5	12.5:5:1
1.0	10	25:5:1
2.0	20	50:5:1
4.0	40	100:5:1

compounds tolerate it). On the other hand, take care that the solutions are at a reaction temperature for a stable interaction equilibrium
- The concentration ratio of substrate to enzyme in this experiment is 5:1 (see Table 11.1). The interaction time of enzyme (respectively, enzyme–inhibitor complex) with substrate in the second tubing (see part 7 in Figure 11.1, length = 830 mm, diameter = 0.25 mm) is 64 s (1.03 min) before the solution reaches the ESI-MS (Q-TOF Ultima API)

11.3.1.2   *Mass Spectrometric Detection*

> Example assay: Monitor the reaction for 120 min. After this time all inhibitor injections are finished (Table 11.1). Note, every concentration is injected twice

11.3.2   Data Analysis

Typically, the manufacturer's analysis software works well for standardized experimental setups but lacks some special features for the evaluation of experimental data from the setup presented here. For instance, studying complex formation with the current experimental setup generates negative chromatographic signals (see NAG_4+H^+ trace in Figure 11.2), which in general can be shown but cannot be used for calculation by manufacturer's software. Thus, we developed and applied a specialized freeware program to allow specialized features and calculations with 'atypical' analytical datasets.

The injection of an inhibitor $((GlcNAc)_3)$ into the flow of the enzyme (HEWL) resulted in an increased mass trace of the inhibitor itself (NAG_3+H^+ trace in Figure 11.2) and a decrease of the product ion trace ($(GlcNAc)_4$; NAG_4+H^+ trace in Figure 11.2). Also the IS [Mal+3H] ion trace is decreasing beginning with 5 nmol injection of NAG_3. Thus the results cannot be directly used quantitatively without correction procedures. However, because the focus in this chapter is elsewhere, the uncorrected data is used, employing a qualitative processing.

Proceed as follows in the manufacturers' software:

- Extract the EIC of substrate, product(s) and inhibitor (ligand) from the mass spectrum. Plot the intensity according to the time points
- Compare the EIC of the product and inhibitor. When the inhibitor plot shows a positive signal, a negative signal should be observed in the product EIC plot and no signal change in the IS EIC plot at the same time point
  > Example assay: The EIC of substrate signal with m/z 1237.4 $[(GlcNAc)_6]$, inhibitor signal with m/z 628.3 $[(GlcNAc)_3]$, and the product signals with m/z 831.3 $[(GlcNAc)_4]$ as well as the signal of the internal standard malantide with m/z 545.3 are plotted in Figure 11.2. The continuous-flow mixing assay shows the expected decreasing product trace while the inhibitor trace is increasing

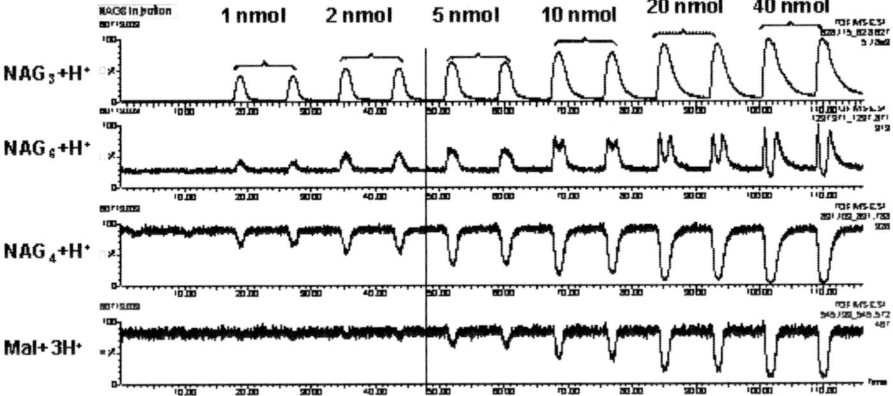

Figure 11.2 EICs of inhibitor ((GlcNAc)$_3$ or NAG$_3$), substrate ((GlcNAc)$_6$ or NAG$_6$), product ((GlcNAc)$_4$ or NAG$_4$) and IS (Mal+3H$^+$) inHEWL assay with (GlcNAc)$_3$ injections, presented with MassLynx software from Waters-Micromass.

(i) Note: The abundances of equal or different signal intensities of the various ion traces do not reflect quantitative data, and signal intensities are not directly comparable between the different ions. The negative tip of IS above 5 nmol inhibitor leads to further need for data processing prior quantitative statements. However, as stated above, here the uncorrected data is applied for the proof of the software tool

(i) The different EIC can be presented by using MassLynx software from Waters for data acquisition (see Figure 11.2). The chromatograms can also be plotted with our Achroma software. As well as the typical positive signals, the negative signals applied in this study can also be integrated and the peak areas calculated

(i) Before using the Achroma software, a data transformation step is necessary. MassLynx, for instance, includes the so-called DataBridge software for converting data files into other formats

- Open the DataBridge software and convert the raw data file into a (MassLynx) ASCII file without header

   (i) An ASCII file without header has the format seen in Table 11.2 and the file extension 'txt'

- The module applied in processing and calculating the enzyme–inhibitor complex EICs is an ion trace comparison. Therefore, open the Achroma software and choose the tool: 'Chromatogram Comparison'. Then load the ASCII data file into the software

   (i) Achroma generates an index file, for faster internal work. The first loading of a data file, *e.g.* with a size of 8 GB, takes approximately 2 min, depending on the computer hardware, because the index file is generated simultaneously during the loading process. Later loadings of this data file do not take so long if the index file still exists.

Table 11.2 Format of an ASCII data file (left) and the presentation in an example file (right)

	Data file format	*Example file (60711Z02.txt)*
Line number		FUNCTION
001	FUNCTION	Scan 1
002	[Empty line]	Retention Time 25
003	Scan[tabulator][tabulator][Number]	199.3016 0
004	Retention Time[tabulator][time]	199.6646 0
005	[Empty line]	199.6704 0
006	[mz value][tabulator][intensity]	...
007	[mz value][tabulator][intensity]	3002.8877 1
...		Scan 2
100	[Empty line]	Retention Time 44
101	Scan[tabulator][tabulator][Number]	199.3016 0

The index file has the same file name as the data file plus an underscore sign at the end of the name (Example file in Table 11.2: 60711Z02.txt, index file name: 60711Z02_.txt) and will be saved on the same place where the data file is saved

- Next, load the first complex EIC into the software and then load the second complex EIC that has to be compared. The software automatically calculates a quotient chromatogram with the data of the first-loaded EIC divided by data of the second-loaded EIC

  ⓘ Such a quotient chromatogram (see the upper traces in Figure 11.3 a–c) is a useful software function for data correction (*e.g.* product with IS; data not shown) or the interpretation of the higher-order complex formation as shown in Figure 11.3

  ⓘ Another task for the calculation of a quotient chromatogram is the following: if the experimental setup has two internal standards which flow with the samples through the system and/or are mixed into the flow before MS, then the quotient of the data of the internal standards should be constant over the complete runtime. The experimental setup works without any influence of system errors, if the quotient chromatogram is constant, whereas fluctuations occur when the flow system shows instabilities

  ➢ Example data analysis: The results of the study of HEWL–inhibitor $(GlcNAc)_3$–complex (LIC) formation by inhibitor flow injections from 1 nmol to 40 nmol (Table 11.1) are shown in Figure 11.3. In each section (A–C) there are always three plots. The upper plot represents a quotient chromatogram calculated from the middle chromatogram plot divided by the lower chromatogram plot as described above. The middle LIC EIC plot in each section has a higher order (n) than the lower LIC EIC plot in the same section. The complexes studied are eightfold charged ions and the ratio enzyme:inhibitor is 1:n, so n = 1–4. The following shortcuts are used for the complexes:

  - LIC1 ($[HEWL((GlcNAc)_3]^{8+}$) for a complex with n = 1
  - LIC2 ($[HEWL((GlcNAc)_3)_2]^{8+}$) for n = 2

- LIC3 ([HEWL((GlcNAc)$_3$)$_3$]$^{8+}$) for n $= 3$
- LIC4 ([HEWL((GlcNAc)$_3$)$_4$]$^{8+}$) for n $= 4$

In Figure 11.3a, the comparison of LIC2 with LIC1 shows that LIC1 is formed above 1 nmol inhibitor injection and LIC2 is formed above 2 nmol. Peak breaks are observed for LIC1 above 5 nmol injections and for LIC2 above 10 nmol injections. The signal of LIC1 has generally a higher intensity than LIC2. However, as stated above, this is not a direct quantitative information, because the ionization efficiency of similar molecules can differ immensely.[9] On the other hand, it can be assumed that LIC1 is formed preferably, because one inhibitor molecule could bind stably at the active site of the enzyme at the lowest inhibitor concentration.

(i) Note, the results in this form give no clear picture of interaction specificities. Specific and unspecific interactions can be identified by further experiments with known ligands or tandem mass spectrometry

The LIC2 is probably the specific interaction of an inhibitor monomer and the active centre of one enzyme molecule and a second unspecific interaction of one inhibitor molecule with the surface of the enzyme (Figure 11.3a). The LIC3 is probably the specific interaction of an inhibitor monomer and the active centre of one enzyme molecule and a second unspecific interaction of an inhibitor homodimer with the surface of the enzyme (Figure 11.3b). Finally, LIC4 is probably the specific interaction of an inhibitor monomer and the active centre of one enzyme molecule and a second unspecific interaction of an inhibitor homodimer and a monomer with the surface of the enzyme (Figure 11.3c). The formation of (GlcNAc)$_3$ homodimers in the gas phase was shown in a previous publication with the general mass spectrometric detection applied here.[10]

The quotient chromatogram always shows negative signals, resulting from the fact that the data of the EIC of LIC1 are the divisor of the quotients with higher intensities and therefore higher abundance. Other statements can be made from the quotient chromatogram: a signal in the quotient chromatogram shows that both LICs at the corresponding abundance have a signal; a plateau on the top/bottom of a quotient chromatogram signal indicates that one LIC chromatogram is near or in the stadium of a signal break; the occurrence of signal break in the quotient chromatogram is a sign for signal breaks in both LICs, thus a higher order complex can occur in this inhibitor concentrations or signal suppression (see also Chapter 2). There are analogous considerations for all other cases (see B and C in Figure 11.3).

Another option of the Achroma software is the discovery and calculation of negative chromatographic peaks.

- In the next section, as an example, negative signal areas of the product trace are calculated at the time points of inhibitor injections. Open a new tool of Achroma software, 'Signal Recognition', and load the data file

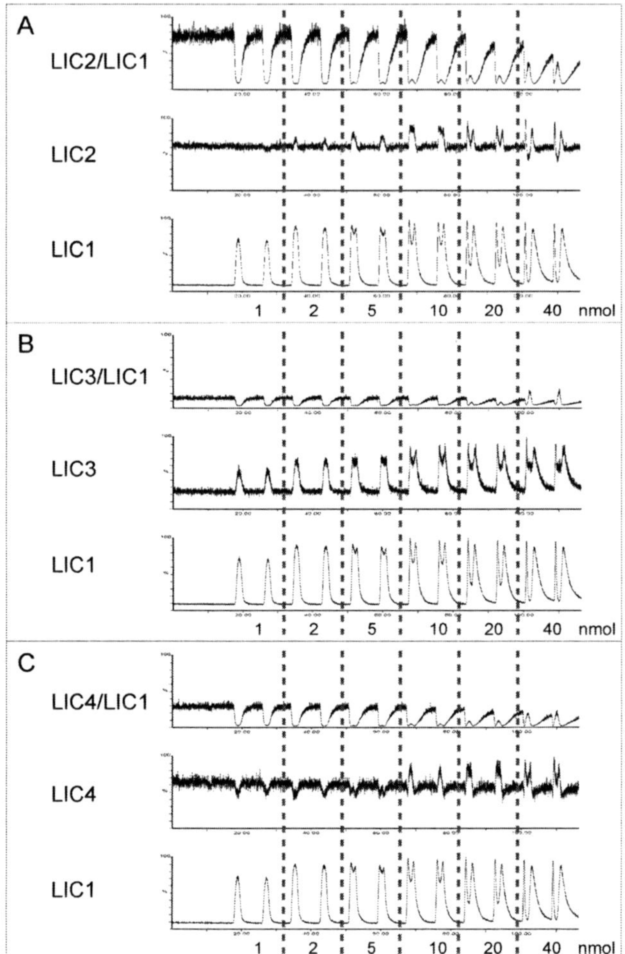

Figure 11.3   Results of data acquisition with Achroma software. Scheme of the plots from A to C: upper plot = quotient chromatogram (middle plot/lower plot), middle plot = complex EIC with higher order than the lower EIC complex plot. (A) quotient chromatogram LIC2/LIC1, LIC2, LIC1; (B) quotient chromatogram LIC3/LIC1, LIC3, LIC1; (C) quotient chromatogram LIC4/LIC1, LIC4, LIC1.

again. After setting the m/z range and choosing the calculation of negative signals, the peak integration can be carried out

> Example assay: Enter m/z range of 831–832, select 'negative signals' into Achroma and then press the 'search' button. The product chromatogram will be shown. After this step press the 'calculate' button and the signal areas will be calculated and be shown in the plot and additionally in a table (Figure 11.4)

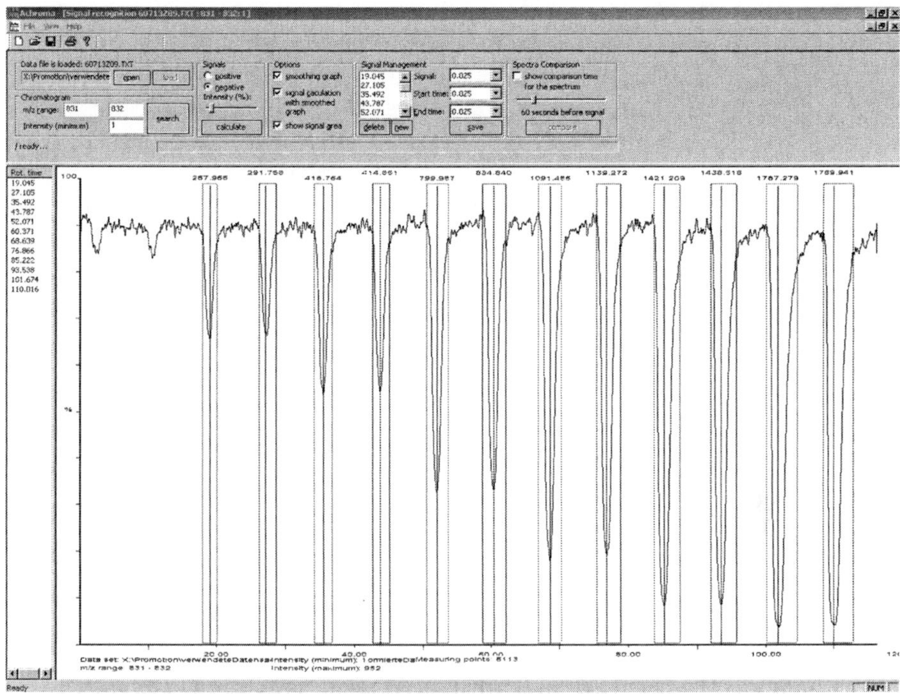

Figure 11.4   Results of the signal recognition with Achroma software tool. Signal areas of the product ($(GlcNAc)_4$) at different inhibitor concentrations (see Table 11.1) are shown.

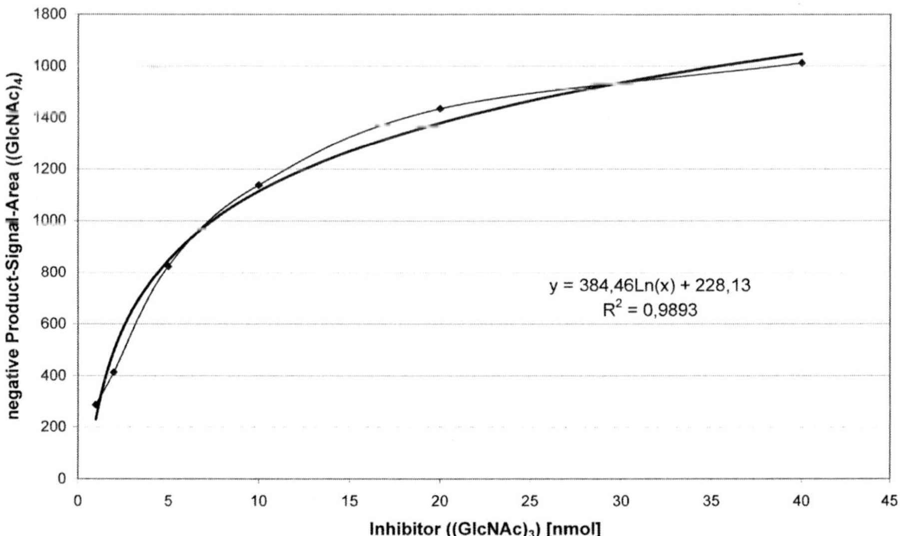

Figure 11.5   Plot of inhibitor against negative product signal areas with trend curve.

- Present a plot: inhibitor concentrations are plotted according to the calculated signal areas of the product
  - Example assay: Product signal areas $(GlcNAc)_4$ *versus* inhibitor concentrations from Table 11.1 are shown in Figure 11.5. It can be shown that the increase of inhibitor abundance does not lead to a linear increase of a 1:1 enzyme–inhibitor complex formation in the mass spectrometer. Several reasons can be ascribed: an equation not established at a specific concentration, a mass spectrometric dissociation, and formation of enzyme–inhibitor complexes with higher inhibitor content (data as shown in Figure 11.3)

11.4 Notes

11.4.1 Working with Enzymes and MS-Based Enzymatic Assays

- See Chapter 10

11.4.2 Working with Achroma External Analytical Software

- The data must have an appropriate format (*e.g.* MassLynx ASCII file without header) to be able to work with Achroma software. If the data file is in a raw data format the Achroma software will not be able to handle it. However, nowadays several raw data formats should be convertible by manufacturer's software into ACSII files or similar data format (txt)
- If the data is in the appropriate format, the analytical results from all commercial mass spectrometers can be used with the software and be combined
- Loading two chromatograms into the module 'Chromatogram comparison' should be done sequentially, because opening/processing two chromatograms simultaneously may lead to software problems, depending on the hardware (processor-intensive calculations)
- For the quotient calculation of two chromatograms the data file may not contain zero values, since division by zero is not defined and will therefore lead to an error
- 'Signal recognition' has an additional feature. It is possible to generate a *difference spectrum* between two time coordinates of the graph (shown in Figure 11.4). For example, subtract the spectrum at the maximum/minimum of a signal and the beginning of a signal and the new *difference spectrum* shows immediately which molecules increase and which molecules decrease in intensity.[5]

At present the Achroma software can received from MK by e-mail. In future it will be downloadable from an Achroma website and is free for non-commercial use.

References

1. T. Niwa, U. Doi, Y. Kato and T. Osawa, *FEBS Lett.*, 1999, **459**, 43.
2. X. Xiao, X Luo, B Chen and S. Yao, *J. Chromatogr.*, 2006, **834**, 48.
3. D. Kolev, *Enzymology*, World Scientific Publishing, NJ, 2007.
4. A. Cornish-Bowden, *Fundamentals of Enzyme Kinetics*, Portland Press, London, 2004.
5. M. Krappmann, T. Letzel, Achroma—a software for analysing 'functional proteomics' research data, like non covalent complex bonding (*e.g.* enzymatic inhibition) got *via* a new LC-MS experimental design, *BMC Bioinformatics*, to be submitted.
6. A. R. deBoer, *Anal. Chem.*, 2004, **76**, 3155.
7. T. Teutenberg, *High-Temperature High-Performance Liquid Chromatography–A User's Guide for Method Development*, Royal Society of Chemistry, Cambridge, 2010.
8. T. Letzel, *Anal. Bioanal. Chem.*, 2008, **390**, 257.
9. N. Dennhart, T. Fukamizo, R. Brzezinski, M. E. Lacombe-Harvey and T. Letzel, *J. Biotechnol.*, 2008, **134**, 253.
10. N. Dennhart and T. Letzel, *Anal. Bioanal. Chem.*, 2006, **386**, 689.

Industrial Standards and Strategies in LC-MS Analysis of Proteins

RENE WISSIACK

Biotech Operations, Process Science, In-Process-Control,
Boehringer Ingelheim RCV GmbH & Co KG, Dr. Boehringer-Gasse 5-11,
1121 Wien, Austria

12.1 Introduction

The determination of average protein molecular weight (MW) in process control plays an important role in the confirmation of protein identity in recombinant protein production. Standard instrumental analytical technologies are well established for the determination of product purity and product concentration at any stage of recombinant protein production.[1,2] Methods may be based on high performance liquid chromatography (HPLC), *e.g.* reversed phase (RP), size exclusion (SEC), protein affinity or ion exchange- (IEX) separation mechanisms (see Chapters 3 and 4) or on gel electrophoresis, *e.g.* sodium dodecyl sulphate polyacrylamide gel electrophoresis (SDS-PAGE), native gel electrophoresis or isolectric focusing (IEF). However, MWs determined by some of these techniques are lacking in accuracy and precision, particularly if average MW for product identification is required or post-translational modification has occurred.[3]

In many cases a more precise MW determination by means of mass spectrometry is required. A significantly higher resolution for intact protein MW measurement is given and the theoretical average MW can be calculated straight

RSC Chromatography Monographs No. 15
Protein and Peptide Analysis by LC-MS: Experimental Strategies
Edited by Thomas Letzel

Published by the Royal Society of Chemistry, www.rsc.org

forwardly;[4] the deviation is directly accessible by mass spectrometry. Reliable results for the average MW are usually obtained within minutes and the determined values can be directly related to calculated theoretical ones. Additionally, coeluting or comigrating protein variants may be identified in many cases.

For instance, the correct cleavage of N-terminal methionine of *E. coli* expressed recombinant proteins is confirmed by measurement of the intact protein MW during all stages of recombinant protein production.[5] While the determination of the correct N-terminus is usually performed by N-terminal Edman degradation, indirect determination by means of electrospray mass spectrometry (ESI-MS) is significantly faster.

However, in general a generic mass spectrometric approach has to be preferred in industry, since samples stemming from different host systems or different protein purification steps have to be analysed within a short period of time. The results obtained might determine the next process step. In order to achieve these goals, rapid and flexible analysis modules are a prerequisite. Possible setups and approaches for fulfilling these requirements are discussed in the presented chapter.

12.2 Materials and Instruments

12.2.1 Reagents, Solvents and Chemicals

- DL-Dithiothreitol, >99%, D5545-5G, Sigma-Aldrich, Steinheim, Germany
- Trifluoroacetic acid (TFA) ULC-MS Optigrade, SO-9668-B001, LGC Promochem, Wesel, Germany
- Acetonitrile, ULC-MS Optigrade, SO-9640-B025, LGC Promochem, Wesel, Germany
- Horse heart myoglobin, >90%, M1882–1G, Sigma-Aldrich, Steinheim, Germany
- Formic acid, pro analysis, 1.00264.1000, Merck, Darmstadt, Germany
- Recombinant proteins with average MWs 10–120 kDa

12.2.2 Analytical Columns and Desalting Cartridges

- MassPREP™ *On-Line* Desalting Cartridges 2.1*10 mm, Waters, Milford, USA
- Microtrap Protein, TR1/25109/03, Michrom Bioresources Inc., Auburn, USA
- ACE C18 300A 3 μm 150*2.1 mm, ACE-211–1502, Bartelt, Graz, Austria
- Other analytical and preparative columns (not mentioned)

12.2.3 HPLC

- Agilent 1100 (1200) series HPLC, Agilent Technologies, Waldbronn, Germany consisting of thermostatted autosampler, thermostatted column oven, binary pump and diode array detector

12.2.4 Preparative Liquid Chromatography

- Äkta Explorer 100 AIR with Fraction Collector, GE Healthcare, Vienna, Austria controlled by Unicorn Software

12.2.5 Mass Spectrometer

- Single quadrupole mass spectrometer 1946 D SL, controlled by Chemstation software revision A.10.02, Deconvolution Software, Agilent Technologies, Waldbronn, Germany
- Quadrupole time-of-flight mass spectrometer Q-Tof Ultima or Ultima II controlled by MassLynx 4.1, MaxEnt1 Deconvolution or BioPharmaLynx 1.2 software, Waters, Milford, US

12.2.6 Direct Infusion

- Syringe pump model 11 Elite, Harvard Apparatus, Hugo Sachs Elektronik, Hugstetten, Germany

12.3 Methods

12.3.1 Mass Spectrometric Detection

- Select the appropriate mass range to detect your protein envelope. Usually set the m/z range as 600–2500 (3000 or even higher)
  - ⓘ Depending on the mass spectrometer type, a smaller scan range and/or slower scanning speeds might be advantageous (*e.g.* quadrupole instruments)
- The parameters of the ion source and the mass spectrometer should be adjusted according to the instrument manual. For further details and hints see Chapter 2
  - ⓘ Parameters such as nebulizer pressure, drying gas flow rates and temperatures are strongly dependent on the composition and applied flow rates of the mobile phase
- Calibrate the mass axis or check the calibrated mass axis in an appropriate manner before measurement
  - ⓘ The mass axis calibration and/or the verification of the mass axis calibration should be performed with independent standards

12.3.2 Protein Mass Spectra Deconvolution

- Software tools for protein mass spectra deconvolution and theoretical MW calculations are available from all mass spectrometer manufacturers. These software tools enable the calculation of the monoisotopic and average MWs of proteins and subsequent comparison with the measured values

- However, deconvoluted protein mass spectra could also be calculated by table calculation programs or deconvolution software such as ESIProt for verification purpose[5,6]
  - ⓘ The MW is calculated in house, avoiding any conflicts over intellectual property right guidelines by divulging the amino acid sequence to outsiders
- The theoretical average MW of a known amino acid sequence (with or without modifications) can be calculated by table calculation programs or open-source tools such as mMass[7,8]
  - ⓘ Average MWs should be carefully checked if they are obtained from unknown sources
- Possible explanations for mass deviations between measured and calculated average MW could be looked up in different databases[9–11]
  - ⓘ More detailed information is available by peptide mapping of the protein after digestion into smaller peptides with proteases, since peptides can be measured more precisely
  - ⓘ Depending on your protein, different proteases (*e.g.* trypsin, chymotrypsin, lysozyme C, or others) might give more advantageous peptides. Check the cleavage sites and determine which protease gives the most promising peptides (0.5–5 kDa)

12.3.3 Intact Protein Analysis by Direct Infusion ESI-MS

- Prepare an appropriate concentrated solution of the protein of interest (*e.g.* 100 mg L^{-1})
- Mix the aqueous protein solution with an acetonitrile/0.1% formic acid solution
  - ⓘ In general the exact concentration of acetonitrile/0.1% formic acid solution is not important, but should be kept constant for one measurement series. Check if a clear solution can be obtained
- Vortex the solution for a couple of seconds, centrifuge for about 30 s
- Direct infusions into the ESI-MS at appropriate flow rates (*i.e.* 2–20 μL min^{-1})
  - ⓘ Choose an appropriate flow rate to obtain a stable spray/signal
  - ⓘ Depending on the signal quality of the spectrum increase or decrease the protein concentration
  - ⓘ Always include a solvent blank measurement to get an idea about chemical noise and contamination of the ion source/mass spectrometer
  - ⓘ Flush the syringe and the capillaries carefully after changing samples and check sample carryover
  - ⓘ Always include a well-known protein standard to check the sensitivity, the resolution and mass axis calibration of the mass spectrometer as a minimum quality assurance

- Ⓘ Some kind of control chart with the average MW of a standard protein should be established
- If suppression of ionization occurs, appropriate desalting of the protein sample solution is required
  - Ⓘ The degree of signal suppression due to salts is depending on the protein
  - Ⓘ Use vacuum centrifugation (for volatile mobile phase modifiers), MW cut-off filters, *off-line* solid phase extraction, dialysis or equivalent to get rid of excess solvents, modifiers and salts

12.3.4 HPLC-ESI-MS

12.3.4.1 On-line Desalting of Proteins by HPLC-MS

- The use of *on-line* desalting cartridges minimizes the sample handling significantly
- 'Peak parking' of the protein during *on-line* or *in-line* preconcentration on a cartridge enables solvent change and desalting for subsequent ESI-MS analysis
  - Ⓘ Polymeric reversed-phase materials at elevated temperatures are a good starting point
  - Ⓘ Elevated column temperatures are preferred
  - Ⓘ Possible gradient: 0–2 min 5% B, 2.01–5 min 30–90% B, 5–7 min 90% B, 7.01–10 min 5% B; solvent A = 0.1% TFA in Milli-Q water, solvent B = 0.1% TFA in acetonitrile
  - Ⓘ If possible, add acetonitrile to solvent bottle A (containing the water) for a protection against rapid algal growth in the bottle
  - Ⓘ Sample introduction can be automated and programmed by parts of the conventional HPLC system and controller software
  - Ⓘ Dilute the sample with solvent A before analysis
- Excessive washing of the cartridge for desalting prior elution and ESI-MS analysis possible
  - Ⓘ The effluent of the cartridge should be switched to waste during the washing step in order to avoid contamination of the ion source
  - Ⓘ Any standard six-port valve of your HPLC system can be used for this type of column switching
  - Ⓘ Forward flushing of the preconcentrated sample during elution offers some kind of additional sample filtration and protects capillaries and sprayer needles of the ESI source
  - Ⓘ Clogged cartridges can be regenerated by reverse flushing, by changing the flow direction on the cartridge
- Mostly, the amount of protein to be injected and analysed is not a limiting factor in recombinant protein product analysis. Increase the concentration and/or injection volume if purified proteins have to be analysed

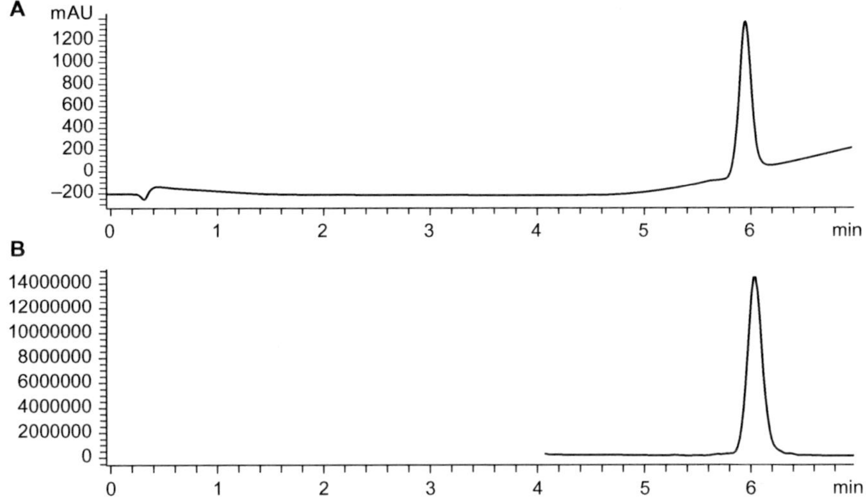

Figure 12.1   HPLC-DAD (A) and HPLC-MS (B) chromatograms of a protein.

- Flow rates and cartridge dimensions are easily adaptable according to protein concentrations, preconcentrated protein solution volumes and the type of mass spectrometer used
  - ⓘ Flow rates for a conventional ion source provided by standard HPLC systems range from 10 µL min^{-1} up to several 100 µL min^{-1}
- Elution of the protein with organic solvent (*e.g.* acetonitrile, methanol, isopropanol)/water mixtures with mobile phase modifiers like formic acid or TFA for subsequent ESI-MS detection
  - ⓘ Usually an efficient desalting step followed by a fast gradient is sufficient to obtain a full separation of salts and proteins, as shown in Figure 12.1
  - ⓘ Memory effects on the cartridge or in the system have to be checked
  - ⓘ Carryover of the autosampler has to be verified. Needle wash programs and blank injection runs must be implemented

12.3.4.2   *RP-HPLC-MS Analysis of Proteins*

- The direct coupling of and RP analytical HPLC column to ESI-MS remains the main problem for this analytical question if maximum sensitivity is demanded.[12] Basically, the drastic decrease in ionization efficiency remains the biggest problem when the favoured ion-pairing reagent TFA is being used as mobile phase additive
  - ⓘ The gain in MS sensitivity is counterbalanced by chromatographic resolution if formic acid or acetic acid is used as a mobile phase modifier instead of TFA

- Since release methods in recombinant protein production are frequently based on RP-HPLC-DAD methods with water/acetonitrile/TFA gradients, direct coupling with these solvents to ESI-MS is preferred
  - ⓘ The chromatographic separation pattern achieved with the HPLC-DAD system remains directly comparable to the HPLC-(DAD)-ESI-MS system

12.3.4.3 Off-line/On-line IEX- or SEC-(HP)LC × HPLC-ESI-MS

- Orthogonal or complementary techniques like SEC, protein affinity chromatography, RP or IEX can be coupled *on-line* to the reversed phase HPLC-MS module as described in section 12.3.4.1 and in Chapter 4
- The peak of interest of SEC-, RP-, protein affinity- or IEX-HPLC can be transferred *on-line* onto the cartridge and subsequently analysed by HPLC-ESI-MS
  - ⓘ If the flow rate of the LC system is too high, collect the fraction of interest, inject a large volume proportion of the solution with the autosampler and preconcentrate it on a desalting cartridge
  - ⓘ Significantly reduced overall analysis time is obtained compared to 'heart cut' of a chromatographic peak followed by lyophilization and reanalysis of the collected fraction

12.3.5 Introducing Selectivity for ESI-MS Protein Analysis by Chemical Modification

12.3.5.1 Non-Reduced versus Fully Reduced Protein Analysis by ESI-MS

- Cysteine links can be reduced (*opened*) and the change in MW compared to intact disulfide bridges (*closed*) could be measured
  - ⓘ The status of the cysteines for recombinant proteins is known and has to be considered for this examination
- The measured differences in average MW corresponds to the number of opened disulfide bridges
  - ⓘ Frequently, denatured proteins (*e.g.* by chaotropic reagents) are not measurable by ESI-MS. At least the average MW of the protein could be determined after additional reduction with DTT increasing signal intensity in ESI-MS
- A significant shift of the charge heterogeneity of proteins to lower mass to charge ratios occurs after denaturing and reducing of disulfide bonds, as demonstrated for recombinant human serum albumin in Figure 12.2
  - ⓘ All amino acids that take up charge are accessible
- Depending on the type of recombinant protein (*e.g.* antibody fragments, dimer, tetramer, light chains, heavy chains) further information about the

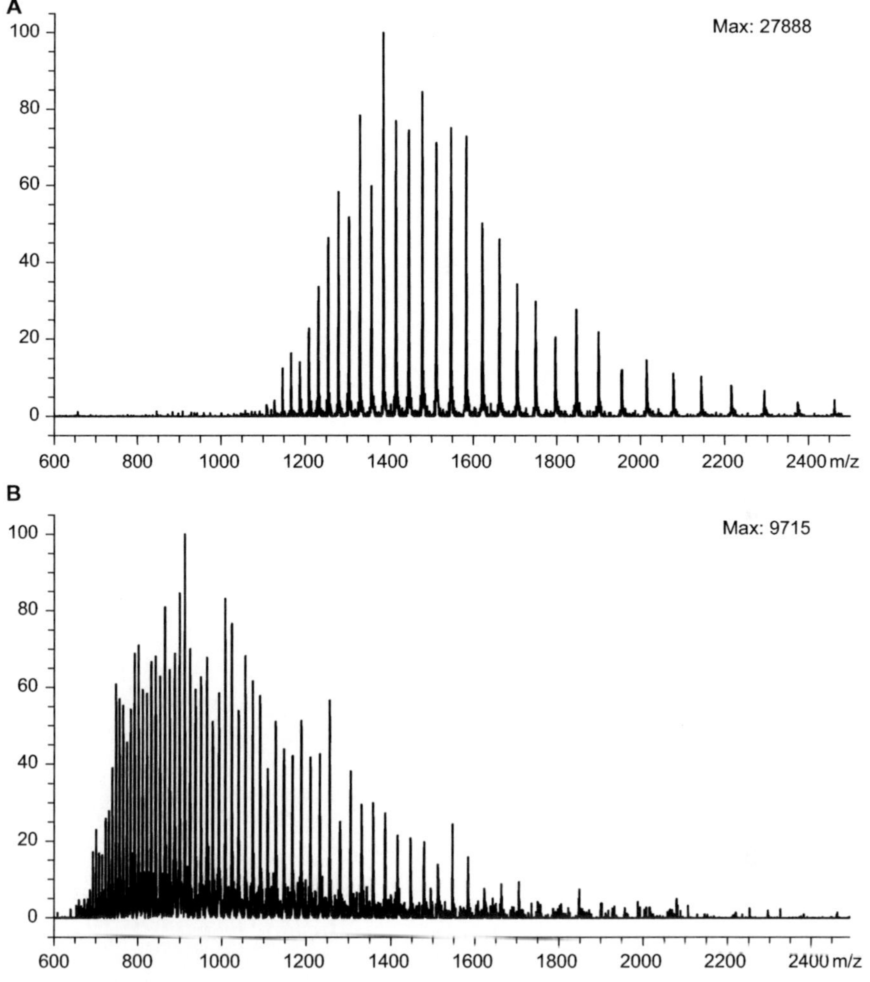

Figure 12.2　Recombinant human serum albumin: intact (A) and reduced/denatured (B).

secondary protein structure is accessible by the analysis of the reduction
products

- ⓘ For example, for antibody fragments the reduction of the intact
  protein results in two smaller proteins differing in retention times as
  shown in Figure 12.3 and in mass spectra caused by a lower average
  MW of the fragments (Figure 12.4)
- ⓘ The smaller proteins are measurable more precisely by means of ESI-
  MS
- ⓘ The non-reduced and reduced protein samples differ only in sample
  preparation; they are measurable within the same analysis method

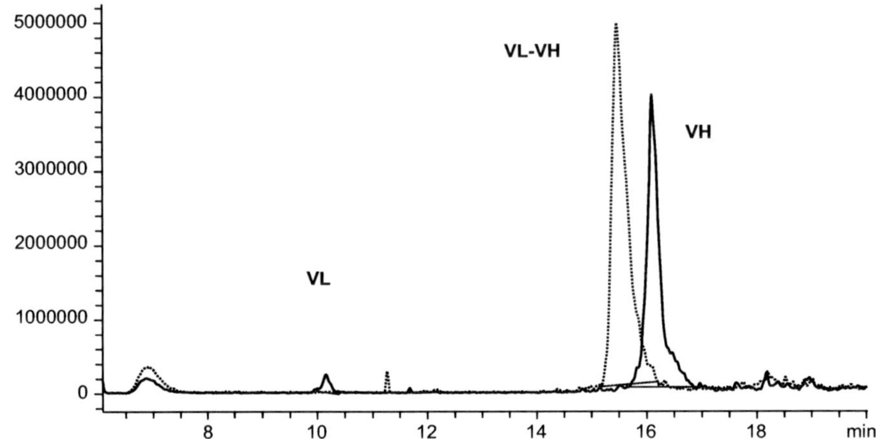

Figure 12.3 RP-HPLC ESI-MS chromatograms: overlay of an intact antibody fragment [VL-VH] and its reduced/denatured chains [VL, VH].

12.3.5.2 Partially Reduced Protein Analysis by ESI-MS

- Frequently, easily accessible cysteines are modified during recombinant protein production. The modification on the formerly free sulfhydryl group of the cysteine can be cleaved by a mild reduction with DTT at room temperature
  - ⓘ Check all process steps to see if any additives are used that could modify/react with the free sulfhydryl groups of cysteines of the recombinant protein during any process step
  - ⓘ The correct average MW of the recombinant protein can be obtained subsequently, tentatively indicating the correct amino acid sequence
- The partial reduction of *intermolecular* disulfide bridges is exploitable for heterodimers, like antibodies, Fab-antibody (fragment of antigen binding) and similar constructs
  - ⓘ Reduction and cleavage of the intermolecular linking disulfide bridge of the heterodimer is rapidly achieved under mild conditions (*e.g.* at room temperature within several minutes by addition of DTT solution), which results in two independent protein chains
  - ⓘ Usually the intramolecular disulfide bridges of the protein chains are not affected under these mild conditions (Figure 12.5)
  - ⓘ If fully reduced protein chains are to be obtained, the reduction can be performed at elevated temperatures as described in section 12.3.5.1
  - ⓘ The reactions can be performed in parallel without significantly increased sample preparation burden and only a few µL of recombinant sample protein solution are required
  - ⓘ The subsequent sample analysis by HPLC-MS without explicit need of baseline separation offers significantly more information about the investigated heterodimer

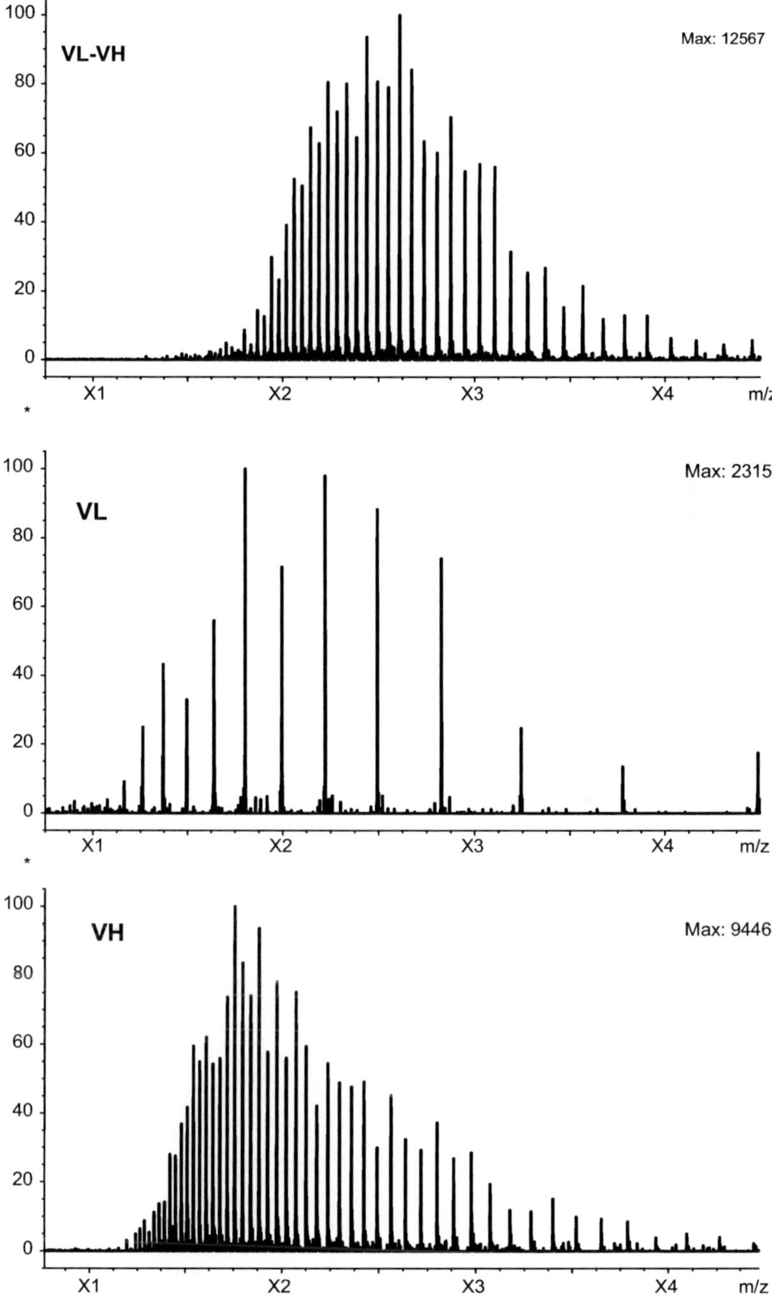

* All mass spectra are given in the same mass axis scale.

Figure 12.4 Mass spectra of an intact antibody fragment [VL-VH] and its reduced/ denatured chains [VL, VH].

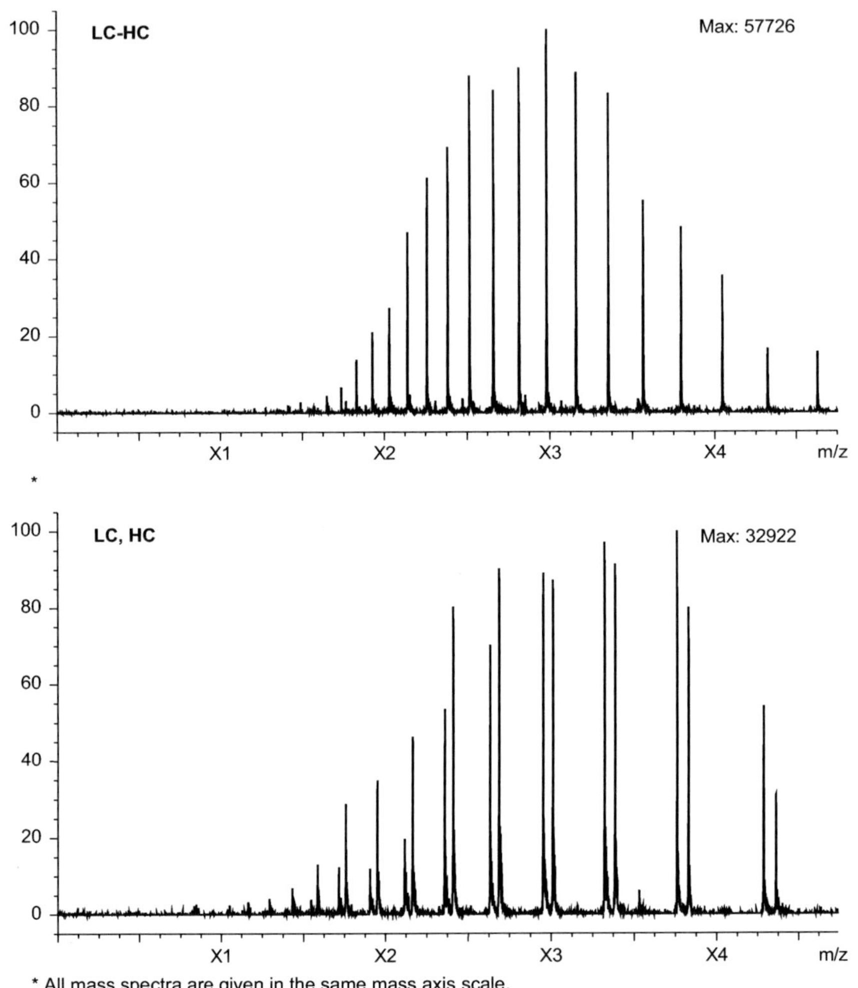

* All mass spectra are given in the same mass axis scale.

Figure 12.5 Mass spectra of a heterodimer Fab antibody [LC-HC] and its individual chains [LC, HC] after selective reduction of the intermolecular linking disulfide bridge.

References

1. F. Lottspeich and J. W. Engels, *Bioanalytik*, Spektrum Akademischer Verlag, Heidelberg, 2006.
2. H. Schlüter H, *Protein liquid chromatography*, ed: Kastner M, Journal of Chromatography Library, Volume 61, Elsevier, Amsterdam, 2000.
3. B. Ardrey, *Liquid Chromatography-Mass Spectrometry: An Introduction, Analytical Techniques in the science*, Wiley, New York, 2003.
4. H. Rehm and T. Letzel, *Der Experimentator Proteinbiochemie/Proteomics*, Spektrum Akademischer Verlag, Heidlleberg, 2009.

5. F. Frotting, A. Martinez, P. Peynot, S. Mitra, R. C. Holz, C. Giglione and T. Meinnel, *Mol. Cell. Proteomics*, 2006, **5.12**, 2336.
6. R. Winkler, *Rapid Commun. Mass Spectrom.*, 2010, **24**, 285.
7. M. Strohalm, D. Kavan, P. Novak, M. Volny and V. Havlicek, *Anal. Chem.*, 2010, **82**, 4648.
8. M. Strohalm, M. Hassman, B. Košata and M. Kodíček, *Rapid Commun. Mass Spec.*, 2008, **22**, 905.
9. Unimod Protein Modifications for Mass Spectrometry: http://www.unimod.org/modifications_list.php?goto=1.
10. Delta Mass, A Database of Protein Post Translational Modifications: http://www.abrf.org/index.cfm/dm.home#top.
11. http://www.expasy.org/tools/#proteome.
12. M. C. Garcia, A. C. Hogenboom, H. Zappey and H. Irth, *J. Chromatogr., B*, 2002, **957**, 187.